美美各其

Embracing the Beauty of All
A Comparison of Beijing Central Axis
and World Urban Axes (Chinese & English)

北京中轴线
与世界都市轴线对比
（汉、英）

吕舟　主编

Chief Editor　LYU Zhou

北京出版集团
北京出版社

各美其美，美美与共

Just as you recognize the merits of your own culture,
so allow different cultures the same recognition.

内蒙古元上都遗址航拍
Aerial view of the Site of Xanadu in Inner Mongolia

日本奈良法隆寺建筑群
The architectural complex of Horyuji Temple, Nara, Japan

柬埔寨吴哥窟及通向四方的道路
Angkor Wat and its roads extending in all directions, Cambodia

法国斯特拉斯堡共和国广场建筑群
The architectural complex of Republic Square, Strasbourg, France

美国华盛顿，国会山与纪念碑东西相望
Washington, D.C., USA, where the Capitol Hill and the Washington Monument face each other from east to west

夜色下的罗马协和大道与梵蒂冈圣彼得大教堂
The Via della Conciliazione and St. Peter's Cathedral in Vatican City under the night sky

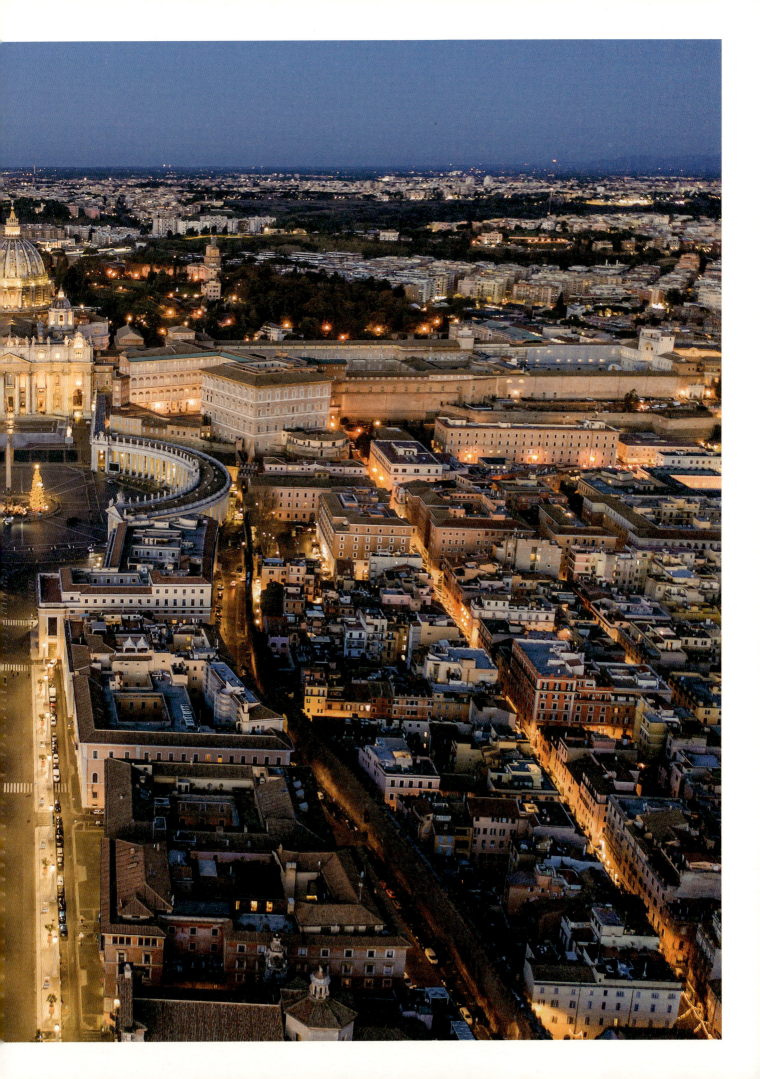

编者说明
Editor's Note

北京是中国现存最完整的古代都城，北京中轴线以其宏大的规模、均衡对称的规划格局、井然有序的城市景观，成为中国传统都城发展至成熟阶段的杰出代表，其深刻体现了中华文明"中"与"和"的哲学理念，强调了中国传统都城规划对秩序的推崇。

2024 年 7 月 27 日，"北京中轴线——中国理想都城秩序的杰作"被成功列入《世界遗产名录》。有趣的是，此次遗产大会的举办地新德里，其城市规划恰好以三角形轴线为"骨架"，建筑风格融合了西方现代和印度传统，形成了独特的"德里模式"。两大东方古国的都市轴线，就这样奇妙地跨越时空相遇了。

《各美其美——北京中轴线与世界都市轴线对比》一书是我们与申遗文本团队再次携手推出的作品。书名"各美其美"一词，取自著名社会学家费孝通先生提出的"各美其美，美人之美，美美与共，天下大同"名句，意指不同国家与民族的文化犹如百花齐放，彼此交流互鉴，共同构成了世界文明图景。

全书分为 6 个部分，第一部分是清华大学教授、北京中轴线申遗文本编制团队负责人吕舟先生撰写的"比较研究视野下的北京中轴线"，梳理了中国传统秩序观念的形成、北京中轴线的历史脉络，并将北京中轴线放在全球范围内，审视其特征与世界遗产价值；第二部分回眸了北京中轴线形成之前的中国都市轴线系统的演进过程；第三至第六部分依次介绍东亚、东南亚、南亚都市的轴线特色，再扩大到全球视角，撷英欧洲、非洲、美洲、大洋洲的一些都市的轴线系统，提供给读者更广阔的视野，从而帮助读者更为深入地理解北京中轴线的突出普遍价值。

本书摄影作品来源于视觉中国、站酷海洛、Veer、图虫等图库，以及波音、王鹤、陈峦、李克君、宋现彬等摄影师；各类都城的轴线示意图由李紫叶绘制。

《天地中和——北京中轴线文化遗产》《各美其美——北京中轴线与世界都市轴线对比》两本书，从酝酿策划到相继出版，历时 5 年有余，其工作涉及方方面面，编者虽孜孜勤勉，但难免会有瑕疵和不足，恳请读者不吝指正。再次向对两本书的出版给予支持和帮助的专家、学者、摄影师以及各方同人致以诚挚谢意！

Beijing is the best-preserved ancient capital city in China. Beijing Central Axis, with its grand scale, balanced and symmetrical layout, and well-ordered urban landscape, stands as an outstanding example representing the mature stage of traditional Chinese capitals. It prominently reflects the philosophical concepts of "centrality" and "harmony" prized in Chinese civilization, highlighting the traditional Chinese capital's emphasis on order in its planning.

On July 27, 2024, "Beijing Central Axis: A Building Ensemble Exhibiting the Ideal Order of the Chinese Capital" was successfully inscribed on the UNESCO *World Heritage List*. Interestingly, the host city of the session of the World Heritage Committee this year, New Delhi, features an urban plan that is structured around a triangular axis, with a unique architectural style blending Western modernism and Indian tradition, creating the distinctive "Delhi Model". In this way, the urban axes of these two ancient Eastern civilizations have wonderfully converged across time and space.

The book *Embracing the Beauty of All: A Comparison of Beijing Central Axis and World Urban Axes* is another collaborative work with the team for the preparation of the nomination dossier for Beijing Central Axis. The title "Embracing the Beauty of All" is inspired by the famous quote from the renowned sociologist Fei Xiaotong: "Just as you recognize the merits of your own culture, so allow different cultures the same recognition." This phrase reflects the idea that the cultures of different nations and peoples are like a blooming garden, engaging in exchange and mutual learning and collectively shaping the tapestry of global civilization.

The book is divided into six sections. The first section is written by Professor Lyu Zhou of Tsinghua University, who is also the head of the World Heritage nomination dossier team for Beijing Central Axis. It is titled "Beijing Central Axis from the Perspective of Comparative Research", where he traces the formation of the traditional Chinese concept of order and the historical context of Beijing Central Axis, and examines its features and World Heritage values from a global perspective. The second section reflects on the evolution of the urban axis systems in China before the formation of Beijing Central Axis. Section three to six introduce the characteristics of urban axes in East Asia, Southeast Asia and South Asia, expanding the scope to a global perspective, where the urban axis systems of cities in Europe, Africa, the Americas, and Oceania are also explored. These sections provide readers with a broader perspective, helping them to gain a deeper understanding of the outstanding universal value of Beijing Central Axis.

Photos in this book are sourced from image libraries such as Visual China, ZCOOL, Veer, and Tuchong, as well as photographers Bo Yin, Wang He, Chen Min, Li Kejun and Song Xianbin. The schematic diagrams of urban axes from various cities were created by Li Ziye.

Heaven and Earth in Harmony: The Cultural Heritage of Beijing Central Axis and *Embracing the Beauty of All: A Comparison of Beijing Central Axis and World Urban Axes* are two books that, from conception and planning to eventual publication, took over five years to complete. The work involved many aspects, and although the editors have worked diligently, there may still be imperfections and shortcomings. We kindly ask readers for their understanding and constructive feedback. We would like to express our sincere gratitude to the experts, scholars, photographers and all colleagues who have supported and contributed to the publication of these two books!

中国北京，从永定门北眺中轴线
Looking north from the Yongdingmen Gate along the Central Axis, Beijing, China

目录 Contents

北京中轴线"景山—钟鼓楼"段落
The Jingshan Hill to the Bell and Drum Towers section of Beijing Central Axis

比较研究视野下的北京中轴线

Beijing Central Axis from the Perspective of Comparative Research

吕舟

LYU Zhou

（北京中轴线申报世界文化遗产文本编制团队负责人、清华大学教授）

(Leader of the Nomination Dossier Team for Beijing Central Axis,
Professor at Tsinghua University)

　　北京是中国漫长历史年代中保存最为完整的古代都城实例。从其现存的历史建筑，城市肌理、结构，城市与周边环境之间的关系，以及丰富的历史文献，仍可清晰地看到这座城市形成、发展、演变的过程，以及它与汉唐长安、洛阳、北宋汴梁等历代都城在规划思想、城市形态等方面的继承关系及创新。

　　北京中轴线作为决定北京整个城市形态的核心建筑群，可以被视为北京城市形态中最为重要、最能够反映基于中国传统哲学和世界观的都城规划思想的物质空间载体。北京中轴线作为一幅由建筑与城市空间构成的历史图卷，凝聚了北京代表的中国古代都城发展高峰时期的核心特征。作为当代中国的首都，20 世纪北京完成了从王朝时期的帝王都城向当代人民城市的转化，北京中轴线同样也反映了这一转化过程中开放的城市形态与传统城市肌理、延续的文化传统与现代规划思想之间的相互作用。北京中轴线的建筑形态、空间关系、景观特征也因此而成了中国历史文化价值的载体，表达了中华文明长期以来的精神追求，见证了中国文明精神不断延续和弘扬的过程，成为人们认识古代中国和当代中国的物质和空间载体。

　　城市是一个文明的发展水平、观念特征的代表性物质遗存。城市的形态反映了国家、民族的思想观念，社会的阶级与阶层关系，以及社会的发展水平。不同文明的差异同样会在城市特别是都城的形态中呈现出来。对北京中轴线特征的理解，需要放到时间的演进过程中，理解北京中轴

注：中英文不完全对应。Please note that the English translation is not the whole text.

北京正阳门
The Zhengyangmen Gate in Beijing

线如何从早期中国建筑的轴线关系和观念中的"中正"，发展到北京中轴线所呈现的精神与物质相互交融、礼乐合一的复杂关系；同时需要将其放到更为广大的人类文明的空间范围上，通过观念的比较，理解北京中轴线的建筑空间形态所呈现的中华文明精神观念对城市和人的生活秩序的塑造。

元代以前中华民族的秩序观念与建筑、城市轴线的发展

"象天法地"是中国传统精神的重要部分，它强调通过对天的运行法则的模仿来构建人间的秩序。这种秩序在中国古代观念中被称为"礼序"，《礼记·礼运》载"故圣人参于天地，并于鬼神，以治政也，处其所存，礼之序也"。"礼"在中国早期文明发展过程中从祭祀的规范演化为行为的规范，并被贯穿到国家治理和生活的行为举止上。如同《左传》（隐公十一年）中所言："礼，经国家，定社稷，序民人，利后嗣者也。"这种观念构建了覆盖整个社会生活的社会角色和与之对应的行为准则之间的关系，这种关系又被视为国家稳定的基础。

早在新石器时代的聚落遗址，如半坡和姜寨遗址，就可以看到处于聚落群组中心的"大房子"，这些大房子影响了周围小房子的布局，从而影响了整个聚落的形态。在被认定为夏、商时期都城的二里头遗址的"宫殿"遗迹中，已有通过轴线来控制整个建筑群形态的痕迹。

公元前1046年周朝建立。在吸收、继承夏商时期礼仪的基础上，周朝形成和完善了一套"郁郁乎文哉"的包含从国家制度、社会生活到城市营建的秩序规定。在关于城市营建规定中包含了城市的大小尺度，这种尺度从国家都城、公爵封邑的方九里，到侯、伯封邑的方七里，再到更低等级封邑的方五里，以及城中道路的宽度、城墙高度等依次递减。都城的核心建筑群则包括了"面

朝后市"的朝堂部分和市井以及"左祖右社"的祖庙(太庙)和国家祭祀的社稷坛。这些内容被成书于春秋战国时期的《考工记》记录下来。在都城核心建筑群的布局方式上,作为国家治理机构的朝堂与民众生活的市井,祭祀祖先的太庙与祭祀国家的社稷坛的空间关系,既反映了传统的"家国一体"的观念,又清晰地表达了它们之间复杂的"礼序"关系。

秦代是中国第一个统一的王朝,它在极短的时间内修筑通往各地的驰道、运河,统一了文字和度量衡,形成了"车同轨,书同文"的景象,不仅完成了国家的统一,更实现了文化的统一。对宏大功业的追求,使得秦代的都城咸阳更呈现了"象天法地"的独特景象,化渭水为银河,将都城中主要的宫殿、政府机构如同夜空中的星座一般排列,而最重要的咸阳宫则与星空中的紫微垣相对应,这种城市布局的形态也影响了汉代都城长安城的布局,有城市史的研究成果认为,汉长安城不规则的城市轮廓对应了紫微垣的形态。

公元前2世纪汉代推崇儒家思想,把儒家思想作为治理国家的基本理念,并对以往的儒家经典文献进行了整理,周代的礼仪制度被作为儒家观念中"礼"的范本。汉代对《周礼》进行了编订,将《考工记》作为《周礼》传播过程中缺失的部分补入《周礼》,成为《周礼》的6个章节之一。这样就使《考工记》中的营城制度成为儒家构建的国家制度的一部分,具有了远超出城市规划的制度意义。

根据城市历史的研究成果,在东汉晚期营建的曹魏邺城出现了贯穿皇城和大城城门的城市轴线。这一轴线由位于轴线顶端的帝王大朝殿宇和从这一大朝殿宇直通城门的道路组成,而道路两侧的政府机构和坊巷建筑呈相对对称的形式布局,形成了城市的主干道。这种轴线形态,一方面是为了帝王的政令能够更快地发布到各地,同时也彰显了帝王的威严。这一都城轴线形态在魏晋时期的洛阳城、东晋和南朝时期的建康城、北魏洛阳城中都得到了延续和发展。

这种城市形态在隋唐时期的都城长安城(隋大兴城)发展到成熟阶段。长安城将帝王的宫殿置于城市的北端,从大朝殿宇所在的太极宫延伸出穿过宫城的承天门、皇城的朱雀门、大城的明德门的城市主干道贯穿整个都城,并在皇城中按照左祖右社的布局方式设置了太庙和社稷坛,在大城中于轴线左右对称的位置设置了东市、西市和规整的街巷、里坊。这种秩序严整的城市形态,也对周边国家产生了影响。日本的平城京和平安京都是长安城规划形态对日本都城影响的典型案例。

这种轴线对称的都城形态,不仅具有空间和建筑布局的意义,更在社会发展过程中被赋予了观念和道德的意义。北宋建立之初,宋太祖赵匡胤修整宫殿,各殿宇与门阙相对,宫殿建成后赵匡胤令臣下打开各殿门,看到整齐相对的建筑形态,赵匡胤对大臣们说,我的心就须如此端正,如果稍有偏差,你们必然能看得清清楚楚。

北京中轴线的历史沿革与形态演进

1267年,忽必烈兴建新的都城——大都(今天北京的前身),这标志着中国古代都城的建设进入一个新的阶段。

根据《析津志辑佚》,忽必烈命刘秉忠营建大都城,刘秉忠先建中心台,以确定这座新都城的中心点,之后与忽必烈确定了城市的朝向,兴建城市重要的宫殿、官署、坛庙和四周城墙,之后通过街巷划分土地,再将土地按职级高低分配给官员,或按出资多少出售给民众,最终完成城市的建设。马可·波罗曾在他的游记中,对他看到的大都城做了这样的描述:"此城之广袤,说如下方:周围有二十四哩,其形正方……全城有十二门,各门之上有一大宫,颇壮丽。四面各有三门五宫,汉代对《周礼》进行了编订，将《考工记》作为《周礼》传播过程中缺失的部分补入《周礼》，成为《周礼》的6个章节之一。这样就使《考工记》中的营城制度成为儒家构建的国家制度的一部分，具有了远超出城市规划的制度意义。

In the Han Dynasty, the *Rites of Zhou* was revised, with *Kaogongji* inserted as a missing section within its transmission. This addition made *Kaogongji* one of the six chapters of the *Rites of Zhou*, thereby incorporating the city planning principles set out by *Kaogongji* into the Confucian framework of state institutions, giving it significance far beyond urban planning.

盖每角亦各有一宫，壮丽相等。宫中有殿广大，其中贮藏守城者之兵杖。街道甚直，此端可见彼端，盖其布置，使此门可由街道远望彼门也……城中有壮丽宫殿，复有美丽邸舍甚多。城之中央有一极大宫殿，中悬大钟一口，夜间若鸣钟三下，则禁止人行。鸣钟以后，除为育儿之妇女或病人之需要外，无人敢通行道中。"

从已有的城市史研究成果看，大都城是秦汉以来第一个完整附会《考工记》"面朝后市""左祖右社"的城市中心建筑群布局形态的都城。之前的都城大多呈现了"左祖右社"的祭祀空间形态，大都则同时遵循了"面朝后市"的朝堂与街市的布局关系。据《析津志辑佚》，大都的米市、面市、段子市、皮帽市、穷汉市、鹅鸭市、柴炭市、铁器市等都在钟楼附近。形成了从中心台到丽正门，涵盖宫廷、朝堂和街市的长度超过 7 华里（约 3.8 千米）的南半城的轴线建筑群，坛庙和社稷坛分别位于中轴线建筑群的东西两侧。这种对《考工记》都城规制的附会，也可从忽必烈 1271 年改国号为"大元"时的《建国号诏》中看到其端倪。在《建国号诏》中，忽必烈将自己比作尧、舜、禹等上古圣王，将自己建立的朝代放到秦、汉、隋、唐的中华文明朝代顺序中，并从《周易》中选择国号。附会《周礼》的都城制度，从《周易》中选择国号，将自己建立的朝代与前朝接续，元代的这一段历史也在一定程度上显示出中华文明的延续性和统一性。

1420 年明代永乐皇帝建成北京城的宫殿、坛庙等重要建筑群，并迁都北京。明代初期元大都城的宫殿建筑群已被拆毁，明代的北京城大体沿用了元代大都城的城址，中轴线建筑群则是在元代建筑的旧址上重新建设的。由于明代放弃了元代相对空旷的北半城，原南半城轴线也随之变成贯穿全城南北，包括北部的钟鼓楼、街市，经皇城、宫城到正阳门的中轴线建筑群。在这一新的中轴线建筑群的规划设计中，元代时位于城市东西两边的太庙和社稷坛被紧密地组织在皇城内，紧邻中轴线的东西两侧，使整个布局更为严谨。这样便形成了永乐时期长度超过 9 华里（约 4.8 千米）的贯穿都城南北的北京中轴线建筑群。同时遵循南郊祭祀的传统，永乐皇帝在北京中轴线南部延长线的东西两侧营建了天地坛和山川坛。显然明代永乐时期的北京中轴线建筑群以及整个城市的营建仍然遵循了《考工记》的都城规划原则，中轴线建筑群严谨符合了"面朝后市""左祖右社"的空间布局。永乐皇帝在其 1420 年发布的《北京宫殿告成诏》中，把这种遵循《考工记》都城建设原则的建设称为"仿古制，循舆情……上以绍皇考太祖高皇帝之先志，下以贻子孙万世之弘规"。

1553 年明代嘉靖皇帝兴建北京外城，修筑永定门，将城南繁华地区及天地坛、山川坛等两组坛庙建筑群包入外城。外城正南门永定门位于北京中轴线上，与内城正南门-正阳门南北相对，

北京天坛祈年殿
The Hall of Prayer for Good Harvests in the Temple of Heaven, Beijing

从体量、形制上都高于东西两侧的城门，而天地坛（后改为天坛）、山川坛（后改为先农坛）巨大的坛域构成了北京中轴线南端东西对称的城市景观。至此，北京中轴线近16华里（7.8千米）的整体规模最终形成。

清代，特别是1750年前后，乾隆皇帝对北京中轴线建筑群做了进一步调整，尤其是对景山的绮望楼、万春亭、观妙亭、周赏亭、富览亭、寿皇殿建筑群等的营建和调整，以及对天桥南侧的景观设计等。这些调整使北京中轴线建筑群的空间节奏、景观特征都得到了强化。

1912年清皇室退位，结束了中国自秦代以来延续了两千年的王朝统治，中国进入了一个新的时代。在这个时代转换变迁中，北京中轴线原来的宫殿、坛庙、园林逐步转变为对公众开放的公园、广场、社会教育设施、博物馆，原本封闭的城市空间也被开放出来成为城市交通通达的区域，并为适应现代城市功能进行了局部改造。例如，天安门广场经历了20世纪50年代和70年代两次重要改建，但这两次改建都遵循了北京中轴线传统的规划思想和理念，使北京中轴线整体的空间环境和景观特征都得到了延续。

在中国古代都城的发展过程中，以中轴线建筑群为代表的城市形态存在一个不断发展和演化的过程，而北京则是这个发展过程最为成熟的案例。它较之其他时期的都城更能完整地体现《考工记》的都城规制，也更典型地反映出以儒家思想为基础的传统文化观念对中国古代都城形态的影响。与其他中国历代都城相比，北京，特别是北京中轴线建筑群是以建筑形态延续至今的唯一案例，其他古代都城，特别是中轴线建筑群在历史发展过程中大多已湮灭消失在城市的演化和建设过程中，存在的部分也仅是以遗址的形态，使人能够追念昔时的风貌。

北京钟楼和鼓楼
The Bell and Drum Towers in Beijing

全球视角下的北京中轴线特征与价值

在判断一种文化是否构成文明的特征时，通常把文字、冶炼技术和是否已形成具有城市形态的聚落作为基本标准。城市之所以能够作为衡量文明发展的标准，是由于它能够反映这个文明社会分工、社会阶层关系等复杂的要素。显然对一座城市，特别是一座都城的研究，也可以从它作为一个特定文明发展的见证的角度去认知它所呈现出的独特性。

北京作为自 13 世纪起元、明、清三个王朝的都城和当代中国的首都，它无疑是中华文明基本特征的载体，是中华民族历史文化价值和精神追求的表达。北京中轴线作为北京超过七个世纪的历史发展的结晶和最具特征的部分，同样也是中华文明的历史文化价值和精神追求的独特见证。

中国古代都城在漫长的历史过程中，作为国家的政治中心，尤其是汉代之后，在尊崇儒家思想的国家治理观念的影响下，城市核心建筑群逐渐形成了以轴线来组织城市布局和空间的城市形态，并经过多个朝代的发展，最终构建起了体现这种传统哲学和世俗社会治理要求的礼仪、秩序。这种礼仪秩序体现为国家各个社会角色的位置、对应的责任和应当遵守的行为准则，并将这种礼仪、秩序体现在北京中轴线的功能分区、空间组织，包括高大建筑、山体与水系、桥梁的关系，建筑的位置、形式、体量、装饰，以及道路的宽度，路幅、铺设方式，植物的种植等各个方面。作为决定整个城市形态的核心建筑群，北京中轴线体现了我国古代社会世俗权力的至高无上，这种至高无上，不仅从宫廷建筑所占据的重要位置中体现出来，而且也从祭祀活动、礼仪的复杂性上表现出来。北京中轴线上的祭祀空间的设置、祭祀活动的礼仪设置同样也围绕着证明世俗权力的正统性而展开，祭祀的对象既包括具有抽象特质的象征天地运行规律的"天道"，也包括象征文明传承的祖先；既有象征文化精神的孔子，也有佛教、道教、民间信仰中的天神、地祇。这些祭祀建筑和祭祀活动，也是中国传统信仰的复杂性和包容性的见证。

从这个角度，北京中轴线这种以世俗权力为核心的、基于中国传统文化与其所强调的礼仪秩序对应的功能完备、规制严整的核心建筑群就与其他以神权为核心的城市中心建筑群存在精神观念和建筑形式、空间形态上的本质差别，同时也对尊崇儒家思想的周边国家的都城形态产生了影响。

在人类文明早期的都城中，如两河流域和早期埃及的城市中，从现有的研究材料看，尚未发现决定整个城市格局的轴线或轴线建筑群。而在这些城市中，特别是在两河流域的都城中，神庙建筑往往呈现出高耸入云，统领整个城市的形态特征。欧洲希腊文明中的城邦国家城市，同样也是以城市守护神的神庙构成城市的核心。希腊代表性城邦雅典的城市形态为：以供奉城市保护神雅典娜的神庙为核心的、高耸在城市上方的卫城统领整座城市。公元前 8 世纪逐渐崛起的罗马的首都罗马城，也仍然在城市发展之初承袭了希腊时代城市的基本特征，围绕罗马城内的 7 座山丘形成了以神庙和宫殿为中心的城市形态，尽管在罗马帝国时期形成了若干由轴线控制的广场，如恺撒广场、图拉真广场等，但这些帝王广场的轴线并未对城市本身的格局形成控制性影响，同时这些广场最重要的建筑也仍然是神殿。

罗马帝国解体之后，欧洲进入中世纪（5—15 世纪），这一时期欧洲的城市大多基于市民经济的发展需要而多呈现放射状、不规则的环形形态。由于基督教的发展，中世纪欧洲城市的核心也往往是教堂和教堂前的广场。南亚和东南亚地区受古印度文化的影响，神庙也同样占据了城市最核心的部位。受到宗教与信仰的影响，南亚和东南亚地区的城市呈现出十字形的轴线格局，而十字形轴线的中心则是神庙。这种城市形态在柬埔寨的吴哥和泰国的阿育他亚城都有反映。这一时期阿拉伯地区的城市则类似欧洲中世纪城市的形态，市场和清真寺是城市的中心。拉丁美洲区域的城市规划理念则具有强烈的宗教象征意义，城市中心的营建展现出神权的核心地位。

在中国，秦、汉、隋、唐作为国土辽阔的统一国家，它们的都城在统一的规划、建设中形成了不同于其他地区文明的都城的形态特征。这种特征的表现之一，就是通过单一和单方向的轴线控制城市的整体格局，并由此而产生了整个城市强烈的秩序性。这一单一轴线的端点则是帝王的大朝殿宇，世俗权力通过这种城市形态得到了强烈的表达。这一时期中国都城的规划理念因隋唐长安城、洛阳城的范例作用，影响到日本、朝鲜半岛等周边地区。

16—18 世纪，欧洲进入了巴洛克和古典主义艺术时期，轴线与园林景观设计手法被用于城市改造之中，以塑造均衡、统一的城市景观。城市中出现了以宏伟纪念物为节点，呈放射网络状的城市轴线系统，如罗马圣彼得大教堂前的城市轴线以及巴黎路易十四时期重要纪念性建筑周边形成的城市轴线。18 世纪后，这种巴洛克和古典主义城市轴线的规划理念不仅在欧洲被普遍接受，而且也随着欧洲国家在其他大陆殖民地的扩张，形成了对非洲、南亚以及美洲新大陆的城市规划的影响。

16 世纪，北京中轴线已形成完整的形态和巨大的规模，成为这一时期明帝国的象征。随着明代（1368—1644 年）儒家哲学思想在东亚地区影响力的扩散，明北京城的规划对尊崇儒学的周边国家都城形态，特别是对城市中轴线及其上建筑群的空间布局产生了直接影响。这种影响在朝鲜半岛的汉阳城、越南的顺化城都有所反映。与巴洛克和古典主义的城市轴线对整个城市影响的有限性和自身形态的开放性不同，尽管北京中轴线决定了整个城市的布局，但其自身的城市空间形态却是相对封闭的、内向的，其通过建筑与空间形态对礼仪秩序进行表达的同时，更通过其封闭性来呈现其崇高地位。

19—20 世纪，欧洲工业化的发展深刻地影响了城市形态的变革，新的城市规划思想不断涌现，现代主义城市实践逐步展开并影响至全球。而这一时期在美洲，如美国的华盛顿、巴西的巴西利亚、澳洲的澳大利亚的首都堪培拉等城市，在规划中通过道路串联起城市中重要的纪念性建筑和政府建筑，形成国家政治和公共生活的核心，而位于这些轴线上的纪念性建筑或公共建筑则成为轴线的视觉中心，同时轴线也联系起城市不同的功能分区。北京中轴线在 20 世纪 50 年代以后的变化中，特别是天安门广场的改建，可以视为在中国传统规划思想的基础上，在追求与历史环境相和谐、尊重历史建筑的高度、体量基础上，结合现代城市规划原则进行的满足当代城市功能的探索。

综上所述，北京中轴线所代表的中国传统都城规划源自其规划理念的儒家哲学和文化背景，代表了东方文明特征的重要方面，在人类城市的发展历史中呈现出了具有独特性的城市形态。北京中轴线作为这一规划理念影响下形成的城市核心建筑群的典范之作，以恢宏的规模、严整的规划格局与建筑群均衡对称的景观形态，全面、充分地展现了中国古代都城规划理念所承载的中华文明突出特征，成为中国传统都城规划理念发展至成熟阶段的杰出范例，并对亚洲周边地区都城规划、营建产生了广泛影响。

Beijing stands as the most well-preserved example of an ancient capital in China's long and storied history. From its surviving historical architecture, urban fabric and layout, to the relationship between the city and its surrounding environment, along with its wealth of historical documents, we can still clearly trace the city's formation, development, and transformation. It also reveals the continuity and innovation in urban planning and design inherited from capitals like Chang'an and Luoyang during the Han and Tang dynasties, and Bianliang of the Northern Song Dynasty.

Beijing Central Axis, as the core architectural ensemble that defines the entire city's layout, can be regarded as the most significant physical embodiment of the capital's planning philosophy, which reflects the traditional Chinese worldview and philosophy. It encapsulates the peak of development for Beijing as an ancient Chinese capital. In the 20th century, as Beijing transitioned from an imperial city of dynasties to a modern metropolis, the Central Axis also mirrored this transformation, where an open urban form interacted with the traditional urban fabric, and cultural continuity blended with modern planning concepts.

河南洛阳，汉魏洛阳故城俯瞰
The Aerial view of the site of Luoyang City from the Han to Northern Wei dynasties, Henan, China

16世纪，北京中轴线已形成完整的形态和巨大的规模，成为这一时期明帝国的象征。随着明代（1368—1644 年）儒家哲学思想在东亚地区影响力的扩散，明北京城的规划对尊崇儒学的周边国家都城形态，特别是对城市中轴线及其上建筑群的空间布局产生了直接影响。

By the 16th century, Beijing Central Axis had already taken on its full form and monumental scale. As Confucianism's influence spread across East Asia during the Ming Dynasty (1368-1644), the planning of Beijing had a direct impact on the capital cities of neighboring Confucian nations, particularly in the spatial layout of their central axes and the buildings along them.

From the perspective of human civilization, a city represents the characteristics of its ideas and level of development by the tangible legacy. Understanding the features of Beijing Central Axis requires situating it within the broader context of time – how it evolved from the axial relationships and concept of "centrality" in early Chinese architecture to the complex interaction of spiritual and physical elements, embodying the unity of ritual and entertainment. At the same time, this understanding must be placed within a global framework of human civilization, recognizing how the architectural forms of Beijing Central Axis reflect the values of Chinese civilization and how these values have shaped the city and the lives of its people.

The Concept of Order and the Development of Urban Axes in Chinese Civilization before the Yuan Dynasty

The principle of "imitating heaven for modeling earth" is a core element of traditional Chinese philosophy. It emphasizes the construction of human order by mirroring the laws of the cosmos. In ancient Chinese thought, this order was referred to as ritual order. Over time, the concept of ritual evolved from rules governing sacrificial practices into a broader code of behavior that permeated state governance and everyday conduct.

As early as the Neolithic period, in settlement sites such as Banpo and Jiangzhai, the presence of a large central house influenced the layout of smaller surrounding houses, shaping the overall form of the settlement. At the Erlitou site, believed to be the capital of the Xia and Shang dynasties, the remnants of palace structures reveal early traces of axial alignment used to organize the spatial layout of the complex.

In 1046 BCE, the Zhou Dynasty was founded. Built on the ritual systems inherited from the Xia and Shang periods, the Zhou Dynasty developed and refined a set of regulations governing everything from state institutions to social life and urban construction. These urban planning regulations, recorded in texts such as the *Kaogongji*, specified the scale of cities and their essential structures, including the imperial court, marketplaces, ancestral temples, and altars for state rituals.

The Qin Dynasty, as China's first unified empire, pursued grand achievements, and its capital, Xianyang, embodied the distinctive vision of "imitating heaven for modeling earth". The Wei River was symbolically transformed into the Milky Way, and the main palaces and government buildings were arranged like constellations in the night sky. The most important of these, the Xianyang Palace, was aligned with the Purple Palace, a key celestial region. This layout later influenced the design of Chang'an, the capital of the Han Dynasty.

In the 2nd century BCE, the Han Dynasty embraced Confucian thought, adopting the Zhou Dynasty's ritual system as the model for Confucian concepts of ritual. The *Rites of Zhou* was revised, with *Kaogongji* inserted as a missing section within its transmission. This addition made *Kaogongji* one of the six chapters of the *Rites of Zhou*, thereby incorporating the city planning principles set out by *Kaogongji* into the Confucian framework of state institutions, giving it significance far beyond urban planning.

According to historical research on cities, by the late Eastern Han period, the Cao Wei capital of Ye city featured a central axis that ran through both the imperial and outer city gates. This axial layout was further developed in cities like Luoyang during the Wei and Jin dynasties, Jiankang in the Eastern Jin and Southern dynasties, and Luoyang under the Northern Wei Dynasty.

The capital city of Chang'an (Daxing City in the Sui Dynasty) reached the zenith of this urban form during the Sui and Tang dynasties. The imperial palace was situated at the northern end of the city, with the city's main thoroughfare extending southward from the Taiji Palace, through the Chengtian Gate in the palace walls, the Zhuque Gate in the imperial city, and the Mingde Gate in the outer city. The ancestral temple and the state altar were arranged within the imperial city in a layout that placed the former to the left and the latter to the right of the main axis, while the outer city was organized with symmetrical marketplaces (East Market and West Market) and a grid of streets and wards. The urban planning of Japan's ancient capitals, such as Heijokyo and Heiankyo, was directly influenced by this design.

The symmetrical, axial layout of capitals not only carried spatial and architectural significance but also came to embody philosophical and moral ideals. When the Northern Song Dynasty was established, Emperor Taizu (Zhao Kuangyin) ordered renovations of the imperial palace. Observing the strict symmetry of the buildings, he remarked to his ministers, "My heart must be as upright as this. If it deviates even slightly, you will clearly see it."

The historical development and morphological evolution of Beijing Central Axis

In 1267, Kublai Khan initiated the construction of a new capital, Dadu (the predecessor of modern-day Beijing), marking a new stage in the development of ancient Chinese capitals. He appointed Liu Bingzhong to oversee the construction. Liu first established the central platform to determine the city's center point. Then, in collaboration with Kublai Khan, the orientation of the city was decided, followed by the construction of important structures such as palaces, government offices, temples, and the surrounding city walls. After that, the land was divided into districts through the planning of streets and alleys and then given or sold to people for building their houses, thus culminating in the completion of the city.

Dadu was the first capital since the Qin and Han dynasties to fully follow the planning principles of the *Kaogongji*, which emphasized the layout of the city with "the palace facing south and the marketplace at the rear", along with the placement of the ancestral temple to the left and altar of land and grain to the right. Earlier capitals typically displayed only the "ancestral temple to the left, the altar of land and grain to the right" layout for ceremonial spaces. In contrast, Dadu also adhered to the principle of having the palace face south and the marketplace at the rear, forming a southern city axis that stretched over 7 *li* (approximately 3.8 kilometers) from the central platform to the Lizhengmen Gate, encompassing palaces, ceremonial halls, and marketplaces. The ancestral temple and the altar of land and grain were positioned on the east and west sides of this central axis respectively.

In 1420, Emperor Yongle of the Ming Dynasty completed the construction of key architectural complexes in Beijing, including palaces and temples, and moved the capital to Beijing. The Ming capital largely followed the site of Yuan Dadu, and the central axis buildings were reconstructed on the foundations of Yuan Dynasty structures. The more open northern half of the city from the Yuan era was abandoned by the Ming, transforming the former southern city axis into a thoroughfare that spanned the entire city from north to south. This new axis included structures like the Bell and Drum Towers in the north, marketplaces, the imperial city, and the palace city, which extended south to the Zhengyangmen Gate. In the design of this new central axis, the ancestral temple and the altar of land and grain, which were originally located on the east and west sides of the city during the Yuan Dynasty, were incorporated more tightly within the imperial city, close to the central axis, giving the layout a more organized and formal structure. This resulted in a central axis that extended over 9 *li* (approximately 4.8 kilometers) through the entirety and dominated the entire urban pattern of the Ming capital.

In 1553, Emperor Jiajing of the Ming Dynasty expanded the city by constructing the outer city and the Yongdingmen Gate. This addition enclosed the bustling southern parts of the city, as well as important temple complexes such as the Temple of Heaven (formerly the Altar of Heaven) and the Altar of Mountains and Rivers (later the Altar of the God of Agriculture). The Yongdingmen Gate, located at the southernmost part of the outer city, was aligned with the central axis, directly opposite Zhengyangmen Gate, the main southern gate of the inner city. The large temple areas created a symmetrical urban landscape at the southern end of the axis. By this time, the central axis had reached nearly 16 *li* (approximately 7.8 kilometers), completing its overall structure.

During the Qing Dynasty, particularly around 1750, Emperor Qianlong further adjusted the central axis, notably through the construction and redesign of the Jingshan architectural complex and the landscape south of the Tianqiao area. These changes enhanced the spatial rhythm and scenic characteristics of the central axis.

In 1912, with the abdication of the Qing royal family, China transitioned into a new era, ending dynastic rule of more than 2,000 years that had persisted since the Qin Dynasty. Many of the former imperial palaces, temples, and gardens along the central axis gradually opened to the public, transforming into parks, squares, educational institutions,

越南顺化皇城
The Imperial City of Hue, Vietnam

and museums. The previously enclosed spaces of the city became accessible, serving as hubs for urban transportation. Partial modifications were made to accommodate modern urban functions, such as the renovation of Tiananmen Square during the 1950s and 1970s. These renovations adhered to traditional planning concepts, ensuring the continuity of the spatial environment and landscape features of the central axis.

Throughout the development of ancient Chinese capitals, the city form represented by the central axis underwent continuous evolution. Beijing is the most mature example of this evolution, more faithfully embodying the city planning regulations outlined in *Kaogongji* compared to other capitals. It also exemplifies the influence of Confucian thought on ancient Chinese urban planning. Unlike other ancient Chinese capitals, most of which have lost their central axis structures over time, Beijing's central axis is the only case that has survived in architectural form. Many other ancient capitals' central axis structures have disappeared during historical transformations, with only archaeological sites remaining to evoke their former grandeur.

Features and Values of Beijing Central Axis from a Global Perspective

Beijing, having served as the capital of the Yuan, Ming, and Qing dynasties since the 13th century, and as the capital of modern China, undoubtedly represents the core features of Chinese civilization and embodies the cultural values and spiritual pursuits of the Chinese nation.

Beijing Central Axis, characterized by its emphasis on secular power and its orderly, functionally complete architectural complex rooted in traditional Chinese culture and its ritual hierarchy, fundamentally differs in both spiritual conception and architectural form from other city centers built around religious authority. Furthermore, this secular power-oriented layout has influenced the capital cities of neighboring countries that revered Confucian thought.

In early human civilizations, such as those in Mesopotamia and ancient Egypt, there has been no evidence of an axis or axial architectural complex that determines the overall urban layout. In these cities, especially in the capitals of Mesopotamia, temple structures often dominated the skyline, controlling the city's overall form. Similarly, in Greek city-states, the temples dedicated to city protector gods were central to the urban layout. For instance, in Athens, the Acropolis, with its temple to the city's guardian god Athena, prominently stood above the city, exerting dominance over its form. Even the capital city of Rome, which began to rise in the 8th century BCE, inherited the basic characteristics of Greek cities, with a temple-centered layout around the seven hills of Rome. While axial plazas such as the Forum of

罗马图拉真广场
Trajan's Forum, Rome

Caesar and Trajan emerged during the Roman Empire, these axes did not significantly influence the overall city layout.

After the fall of the Roman Empire, Europe entered the Middle Ages (5th-15th centuries). During this period, European cities, driven by the development of civic economies, often displayed irregular, radial layouts. Due to the rise of Christianity, the core of medieval European cities typically consisted of cathedrals and plazas in front of them. In South Asia and Southeast Asia, influenced by ancient Indian culture, temples similarly occupied the most central positions in cities. Cities in these regions often adopted a cruciform axial layout, with temples at the center, as seen in Angkor in Cambodia and Ayutthaya in Thailand. In the Arab world, cities resembled those in medieval Europe, with markets and mosques serving as the urban centers. In Latin America, city planning also exhibited strong religious symbolism, with urban layouts reflecting the central role of religious authority.

In China, under the unified governance of dynasties like the Qin, Han, Sui, and Tang, capital cities developed a unique form distinct from other civilizations. One key feature of this form was the use of a single, unidirectional axis to control the overall city layout, creating a strong sense of order. The focal point of this axis was the emperor's grand hall, where secular power was prominently expressed through the city's design. The planning concepts of Chinese capitals during this period, exemplified by the cities of Chang'an and Luoyang in the Sui and Tang dynasties, influenced capital designs in neighboring regions like Japan and the Korean Peninsula.

From the 16th to the 18th century, Europe experienced the Baroque and Classical periods, during which axial and streetscape design techniques were applied to urban renovation, creating balanced, unified cityscapes. This period saw the emergence of urban axes organized around monumental nodes, forming radial networks, such as the axis in front of St. Peter's Cathedral in Rome and the urban axes surrounding significant monuments in Paris during the reign of Louis XIV. In the 18th century and beyond, these Baroque and Classical city-planning concepts were widely adopted in Europe and extended to Africa, South Asia, and the Americas through colonial expansion.

By the 16th century, Beijing Central Axis had already taken on its full form and monumental scale. As Confucianism's influence spread across East Asia during the Ming Dynasty (1368-1644), the planning of Beijing had a direct impact on the capital cities of neighboring Confucian nations, particularly in the spatial layout of their central axes and the buildings along them. This influence is evident in the capitals of Hanyang (in Korea) and Hue (in Vietnam). Unlike the open, expansive urban axes of Baroque and Classical cities, which had a limited influence on the city's overall structure, Beijing Central Axis determined the city's entire layout. However, its urban space was relatively enclosed and inward-looking, expressing its supreme status through both its architectural forms and its closed-off nature, while symbolizing ritual order.

In the 19th and 20th centuries, European industrialization profoundly transformed urban forms, and new urban planning ideas emerged. Modernist urban practices gradually took hold and influenced cities worldwide. During this period, cities like Washington, D.C. in the United States, Brasília in Brazil, and Canberra in Australia connected important monumental and governmental buildings through axial roads, forming the core of national political and public life. The monumental or public buildings along these axes became the visual focus of the city, and the axes connected different functional areas of the city. The changes to Beijing Central Axis after the 1950s, particularly the renovation of Tiananmen Square, can be seen as explorations that combined modern urban planning principles with traditional Chinese planning philosophies, respecting the heights and scale of historical buildings while meeting the needs of contemporary urban functions.

In summary, Beijing Central Axis, which represents the traditional Chinese capital planning rooted in Confucian philosophy and cultural context, highlights important aspects of Eastern civilization. In the history of urban development, it presents a unique urban form. As a quintessential example of a city's core architectural ensemble shaped by this planning philosophy, Beijing Central Axis, with its grand scale, orderly layout, and balanced, symmetrical landscape, fully and comprehensively showcases the outstanding features of Chinese civilization embodied in ancient capital planning. It stands as an exemplary case of traditional Chinese capital planning reaching maturity which has exerted widespread influence on capital planning and construction in surrounding regions across Asia.

参考资料References：

1. 《论语·八佾》。
   The Analects, Chapter "Ba Yi".

2. 郭黛姮主编：《中国古代建筑史 第三卷》，中国建筑工业出版社，2003，第19页。
   Guo Daiheng (eds.), *History of Ancient Chinese Architecture, Volume 3*, China Architecture & Building Press, 2003, P.19.

3. 沙海昂注：《马可·波罗行纪》，冯承钧译，冯承钧著译集，上海古籍出版社(Kindle版，位置3234—3236)。
   Chahine S. (annotated), *The Travels of Marco Polo*, Feng Chengjun (trans.), Feng Chengjun Translation Collection, Shanghai Classics Press (Kindle Edition, location 3234-3236).

源远流长

Timeless Legacies

回眸北京中轴线之前的中国都市轴线

A Glance at China's Urban Axes Prior
to Beijing Central Axis

曹魏邺北城金凤台
Jinfeng Terrace, Yebei City of the Cao Wei period

汉魏洛阳故城遗址
Site of Luoyang City from the Han and Northern Wei dynasties

西安大雁塔
Giant Wild Goose Pagoda in Xi'an

开封州桥遗址
Site of Zhouqiao Bridge in Kaifeng

内蒙古正蓝旗元上都遗址
Site of Xanadu in in Zhenglan Banner, Inner Mongolia

"宅兹中国"，青铜器何尊上的这句铭文是"中国"一词最早的文献表述，向我们揭示出早在商周更替的时代，"中"就是一个神圣的信仰概念，它不仅是当时古人信仰体系的核心概念之一，还是规范古人日常生活的准则之一，甚至建城设都皆要符合"中"的标准。

周代将逐渐完善并广播天下的礼制与"以中为尊"的古老信仰结合，使中国古人很早就在建筑和城市营造中有意识地使用轴线和对称，来彰显自己对于国家秩序、礼仪规范和神圣信仰的追求。成书于战国时代的《考工记》为中国理想都城规划范式提供了更加详细的指导方案。虽然从考古学上看，早在约3700年前的二里头时代的宫城就有明显的中轴线道路，但《考工记》所描绘的都城规划，至今尚未发现与之完全相符的早期城址。

中国古代都市轴线经历了漫长的演变历史。当秦始皇统一天下时，其都城仿星汉而建，此后的汉朝都城虽然有棋盘式的网格布局，却也并未有意突出某条轴线。根据目前的考古资料，那种贯穿全城且具有唯一性的都城中轴线，最早雏形出现在曹魏的邺北城，并且直到北魏洛阳城的营建，中轴线才正式确立于城中。从魏晋到隋唐，古代都城逐步出现了单一宫城、三重城规划格局和封闭的棋盘式里坊布局。此后到了北宋的东京城，市井经济活跃，城市街道布局变得开放起来，商贸性的街市出现在都城的中轴线上。

在这长达千年的演变中，《考工记》那种理想都城的规划范式一点点地从文字变成了现实。比如，城内建筑以宫城正殿为基点，向四方铺展；中轴线多自北向南贯穿全城，主轴街道对全城结构整体或局部起到统领性作用；城内重要的官署部门、礼仪祭祀建筑以中轴线对称布局，道路网络、城郭结构基本上也以中轴线为准，左右对称布局。在中轴线概念的统领下，古代中国都城的建筑布局、建筑形制、景观视觉效果越来越规整庄重，营造出层次感和秩序感的空间氛围，突出了儒家礼仪性的追求和皇权至上的威严。

因此，元明清时代的北京中轴线并非凭空而来，它是古代中国几千年都市营造演化历史的结晶，有源远流长且丰富多彩的"前传"。

The phrase "Residing in this Middle Kingdom', inscribed on a bronze vessel, is the earliest documented reference to the term "China". This inscription reveals that as early as the transitional period between the Shang and Zhou dynasties, "Zhong" (Middle) was already a sacred concept. It not only stood as a core belief in the ancient system of faith but also served as one of the principles guiding the daily lives of the ancients. Even the construction of cities and capitals had to align with the standards of "Zhong". *Kaogongji*, written during the Warring States Period, provided a more detailed guide for the planning of an ideal Chinese capital.

The axial layout in ancient Chinese cities underwent a long evolution. According to current archaeological evidence, the earliest prototype of a single, unified central axis traversing the entire city appeared in Cao Wei's Yebei City, but it wasn't until the construction of Northern Wei's Luoyang City that the central axis was formally established within the city. From the Wei and Jin periods to the Sui and Tang dynasties, ancient capitals gradually adopted the single palace city, triple city layout, and grid streets system. By the Northern Song period in Dongjing (modern Kaifeng), as the urban economy thrived, city street layouts became more open, with commercial streets appearing along the central axis of the capital.

Over this thousand-year evolution, the planning model of the ideal city as envisioned in *Kaogongji* gradually transitioned from concept to reality. For example, the buildings within the city were arranged with the main hall of the palace city as the starting point, and the streets were planned as a chessboard. The central axis typically ran from north to south, traversing the entire city, with the main streets playing a commanding role over the city's overall or partial structure. Important government offices, ritual buildings, and ceremonial spaces were symmetrically arranged along the central axis, and the road network and city fortifications were also designed symmetrically around it.

Thus, the central axis of Beijing during the Yuan, Ming, and Qing dynasties did not emerge out of thin air; it is the culmination of thousands of years of urban construction and evolution in ancient China.

曹魏邺北城
Yebei City of the Cao Wei Period

邺北城遗址区
The Site of Yebei City

中国古代都城轴线演变的不同阶段

The Evolution of Axes in Ancient Chinese Capitals

根据中国古代都城遗址的考古与研究，都城轴线的发展演变大致分为四个阶段：

1.东周以前（公元前770年以前）：都城松散布局中存在由重要建筑对位关系构成的宫殿区轴线。

2.东周至两汉（公元前770—220年）：多宫制格局下以宫城正殿为基点纵贯城市局部区域的轴线。

3.魏晋至隋唐时期（220—907年）：三重城、单一宫城都城格局中，出现以宫城正殿为基点，以中央大街为标志，两侧存在严格对称里坊格局的中轴线。

4.宋至明清时期（960—1911年）：开放街巷格局下的都城中轴线。

Based on archaeological findings and research, the development of capital city axial lines can be roughly divided into four stages:

1. Pre-Eastern Zhou Period (before 770 BCE): In the loosely arranged layouts of early capitals, axial lines were formed within the palace districts through the spatial alignment of significant buildings.

2. The Eastern Zhou to the Han Dynasty (770 BCE-220 CE): During this period, in multi-palace layouts, the axial line extended across part of the city, with the main palace serving as the reference point.

3. The Wei, Jin, Sui, and Tang periods (220-907 CE): In the urban structure of three-tiered cities with a singular palace city, an axial line emerged, centered on the main hall of the palace. This line was marked by a central avenue, with symmetrical city blocks (li-fang) on either side.

4. The Song to Ming-Qing periods (960 CE-1911): In the more open street network layouts of later capitals, a central axial line continued to be a defining feature of urban planning.

曹魏邺北城轴线示意图
Diagram of the axis of Yebei City of the Cao Wei period

邺北城遗址位于今河北省临漳县西南，营建于东汉末年，是曹魏政权的国都。

不同于秦汉时期宫殿散布城内的布局，邺北城把城内的宫殿集中在一起，形成了单一宫城，进而构成宫城、皇城（各官署部门）、郭城的三重都城规划格局。宫城位于全城的北部中央，宫城东西两侧是贵族居住的"戚里"和皇家园林。宫城被一条南北轴线分成两部分，西侧部分位于全城的北部中央，以正殿太极殿为基点，建设了一条纵贯宫城正门、郭城正门的中轴线大街，形成全城的中轴线。考古证实，这条大街宽17米，正对宫殿区主要宫殿，是邺北城的南北主干大道。

邺北城这种都城新格局，对此后中原王朝和周边地区的都城规划均产生了深刻影响。

Built at the end of the Eastern Han Dynasty, Yebei served as the capital of the Cao Wei regime. Yebei concentrated the palace buildings within a walled area, forming a singular palace city, which in turn established a threefold urban planning structure consisting of the palace city, the imperial city, and the outer city. The palace city was located in the northern central part of the city, with aristocratic residences and royal gardens on its east and west sides. The palace city was divided into two parts by a north-south axis. The western section, located at the northern center of the entire city, used the main hall, Taiji Hall, as a reference point, creating a central axis avenue running through the main gates of both the palace city and the outer city. This new layout of the capital city had a profound impact on the urban planning of later Central Plains dynasties and surrounding regions.

北魏洛阳城
Luoyang City of the Northern Wei Dynasty

汉魏洛阳故城宫城遗址
Palace city site of Luoyang from the Han and Wei periods

中轴视角之北魏洛阳城
Northern Wei's Luoyang City from the Central Axis Perspective

　　都城虽然有明显的中轴线并围绕轴线进行了布局，但北魏皇室在规划城市格局时，并未严格遵循《考工记》所载"面朝后市""左祖右社"等布局原则，没有明显的商业街道，礼仪和祭祀建筑也建在郭城之外，并未纳入城市之中。

Although the capital had a clear central axis and its layout was organized around this axis, the Northern Wei's royal family did not strictly follow the urban planning principles recorded in *Kaogongji*, such as the "palace in the front and market at the rear" or the "ancestral temple on the left and altar of land and grain on the right" when designing the city. There were no distinctly defined commercial streets, and the ceremonial and sacrificial buildings were constructed outside the outer city, rather than being incorporated within the urban fabric.

汉魏洛阳故城残垣
Ruins of Luoyang City of the Han and Wei periods

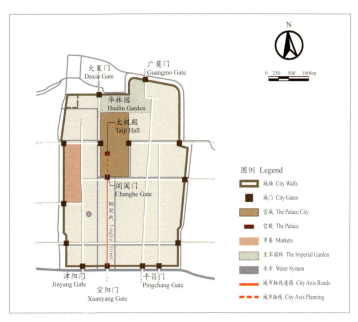

北魏洛阳城轴线示意图
Diagram of the axis of Luoyang City of the Northern Wei Dynasty

　　北魏太和十七年（493年），孝文帝把都城从平城迁至洛阳，在原有汉魏洛阳城的基础上进行改造，把原来的外城化为内城，又在外面另起郭城，加上内城里中北部的宫城，形成了宫城、内城、郭城三重且环环相套的都城布局，此即北魏洛阳城。

　　从宫城正殿太极殿向南通到宫城南门阊阖门的道路，再以中央大街即铜驼街的形式通达内城的南门宣阳门，从而形成了洛阳城的中轴线，中央官署分列于这条中轴线的两侧，城内一些主要街道里坊也左右分布于中轴线两侧，每个里坊四面设门，管理民众进出。在郭城之外的南郊，建设了灵台、明堂等礼仪建筑，分布在中轴线延长线的东侧。

In 493, Emperor Xiaowen relocated the capital from Pingcheng to Luoyang. The former outer city was converted into the inner city, and a new outer city was built around it. Together with the palace city in the central-northern part of the inner city, this created a three-tiered, nested capital layout, known as Northern Wei's Luoyang. The road extended from the main hall of the palace city, southward to the south gate of the palace city, and then continued as the central avenue, reaching the south gate of the inner city. This established the central axis of Luoyang, with the central government offices lined up along both sides of the axis. Several major streets and residential blocks (*li-fang*) in the city were symmetrically distributed on either side of this axis, and each block was enclosed with gates to regulate the entry and exit of residents. In the southern suburbs, beyond the outer city, ceremonial buildings were constructed, located on the east side of the extended central axis.

隋唐长安城
Chang'an City of the Sui and Tang Dynasties

西安大明宫遗址鸟瞰
Aerial view of the Daming Palace site, Xi'an

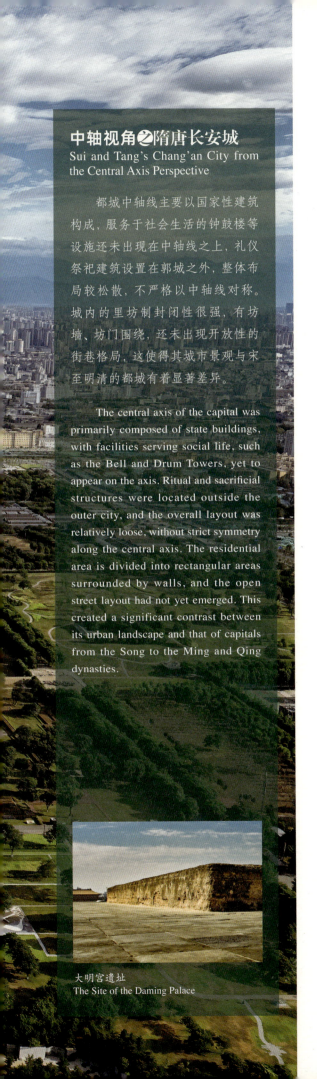

中轴视角之隋唐长安城
Sui and Tang's Chang'an City from the Central Axis Perspective

都城中轴线主要以国家性建筑构成，服务于社会生活的钟鼓楼等设施还未出现在中轴线之上，礼仪祭祀建筑设置在郭城之外，整体布局较松散，不严格以中轴线对称。城内的里坊制封闭性很强，有坊墙、坊门围绕，还未出现开放性的街巷格局，这使得其城市景观与宋至明清的都城有着显著差异。

The central axis of the capital was primarily composed of state buildings, with facilities serving social life, such as the Bell and Drum Towers, yet to appear on the axis. Ritual and sacrificial structures were located outside the outer city, and the overall layout was relatively loose, without strict symmetry along the central axis. The residential area is divided into rectangular areas surrounded by walls, and the open street layout had not yet emerged. This created a significant contrast between its urban landscape and that of capitals from the Song to the Ming and Qing dynasties.

大明宫遗址
The Site of the Daming Palace

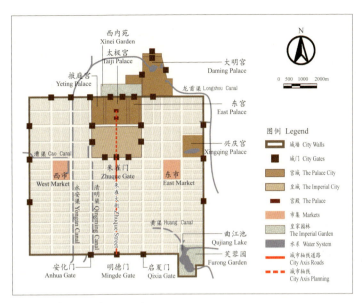

隋唐长安城轴线示意图
Diagram of the axis of Chang'an City of the Sui and Tang dynasties

隋大兴、唐长安城遗址位于今陕西省西安市，兴建于隋文帝开皇二年（582年），是隋唐两代的都城。这座全新规划的都城平面呈近正方形，继承了此前都城的宫城、皇城和郭城三重格局，但宫城位于都城正北部，宫城南侧建皇城，集中设置各官署部门。皇家御苑大明宫位于城外北侧偏东。

长安城的中轴线即南北向的朱雀大街位于都城正中，且与宫城、皇城的轴线重合。在皇城的东南和西南角，沿着中轴线对称设立了宗庙和社稷，符合《考工记》"左祖右社"之制。郭城东、西、南三面各开三座城门。整个长安城的城郭结构、城门分布、街巷里坊、东西两市均严格以中轴线对称分布，可见中轴线在都城规划中无可比拟的重要性。

The site of Sui's Daxing and Tang's Chang'an is located in present-day Xi'an, Shaanxi Province. Constructed in 582, it served as the capital during the Sui and Tang dynasties. The newly planned city had a nearly square layout, and the palace city was situated in the northern part of the capital, with the imperial city built to its south, where various government departments were centralized. The Daming Palace was located outside the city, to the northeast and on the top of Longshouyuan hill. The city's central axis, from the main palace of the palace city through the Zhuque Avenue, was situated in the center of the capital at the southeastern and southwestern areas of the imperial city, ancestral temples and altars were symmetrically positioned along the central axis. The entire layout of Chang'an was strictly symmetrically arranged along the central axis, underscoring its unparalleled importance in the city's planning.

北宋东京城
Dongjing City of the Northern Song Dynasty

州桥遗址。州桥曾是北宋东京城中轴线上的一座重要桥梁
The site of Zhouqiao Bridge, an important bridge located on the central axis of Northern Song's capital, Dongjing

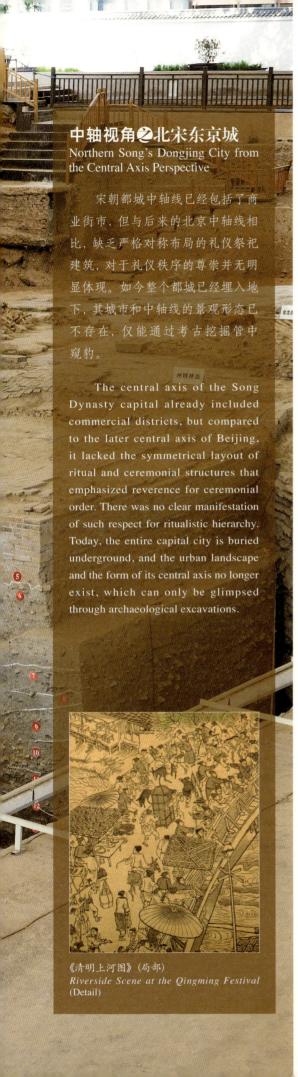

中轴视角之北宋东京城
Northern Song's Dongjing City from the Central Axis Perspective

　　宋朝都城中轴线已经包括了商业街市，但与后来的北京中轴线相比，缺乏严格对称布局的礼仪祭祀建筑，对于礼仪秩序的尊崇并无明显体现。如今整个都城已经埋入地下，其城市和中轴线的景观形态已不存在，仅能通过考古挖掘管中窥豹。

The central axis of the Song Dynasty capital already included commercial districts, but compared to the later central axis of Beijing, it lacked the symmetrical layout of ritual and ceremonial structures that emphasized reverence for ceremonial order. There was no clear manifestation of such respect for ritualistic hierarchy. Today, the entire capital city is buried underground, and the urban landscape and the form of its central axis no longer exist, which can only be glimpsed through archaeological excavations.

《清明上河图》（局部）
Riverside Scene at the Qingming Festival (Detail)

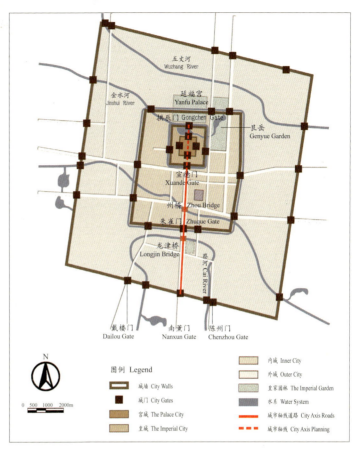

北宋东京城轴线示意图
Diagram of the axis of Dongjing City of the Northern Song Dynasty

　　北宋东京城城址位于今河南省开封市，因黄河泛滥带来泥沙，原城已深埋于地下。

　　根据记载，东京城为宫城、皇城、内城、外城四重城郭环套式布局，其中宫城居中。城内有一条南北方向的御街，此为全城的中轴线。中轴线从宫城出发，向南经过皇城正南门，过古州桥、内城正南门朱雀门、龙津桥，至外城正南门南薰门为止，全长约4.25千米。根据文献记载和考古勘察，东京城内水网纵横，街巷布局较为开放，打破了此前都城里坊制的布局，水系于商业街市应该分布在中轴线沿线的街巷里，使整个城市更有经济活力和生活气息。

Dongjing followed a concentric design with four layers of city walls: the palace city, the imperial city, the inner city, and the outer city, with the palace city at the center. A north-south Imperial Street served as the central axis of the city, stretching approximately 4.25 kilometers. The city was crisscrossed by a network of waterways, and its streets and alleys were laid out in an open fashion, breaking away from the previous urban grid model. The commercial districts were likely distributed along the streets and alleys, infusing the entire city with economic vitality and a lively atmosphere.

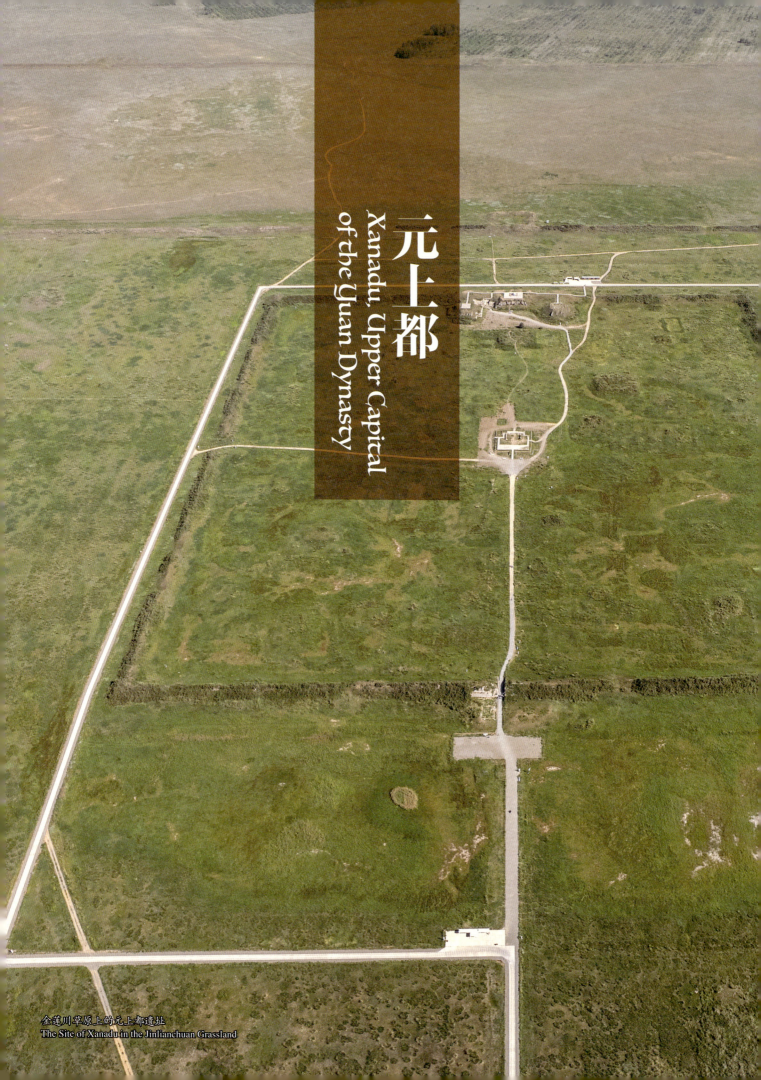

元上都
Xanadu, Upper Capital of the Yuan Dynasty

金莲川草原上的元上都遗址
The Site of Xanadu in the Jinlianchuan Grassland

中轴视角之元上都
Yuan's Xanadu from the Central Axis Perspective

元上都建于元大都之前，与后者规整的格局相比，元上都位于草原地带，建设者因地制宜，更多地保留了游牧民族的社会特点，并未严格形成环套式的城市布局，城市主轴线位于城东侧而非正中。限于考古工作成果，元上都城内礼仪祭祀建筑的分布尚不清晰。

Xanadu was built before Dadu. Compared to the orderly layout of the latter, Xanadu, located in the grassland region, was constructed in harmony with the local environment, preserving more of the nomadic society's characteristics. It did not adopt a strict ring-like city layout, and the main axis of the city was located on the east side rather than in the center. Due to the limited results of archaeological work, the distribution of ceremonial and sacrificial buildings within Xanadu remains unclear.

出土于元上都的汉白玉雕龙角柱
White Marble Corner Pillar with Carved Dragon Pattern, unearthed from the Site of Xanadu

元上都轴线示意图
Diagram of the axis of Xanadu

元上都城址位于今内蒙古自治区正蓝旗，于1256年开始营建，1263年设为上都，与南面的元大都同为元朝的政治中心。

上都城平面为正方形，最外层为外城，外城东南部分为皇城，而宫城位于皇城内中部偏北。其中宫城和皇城有一条南北方向的中轴线，它以宫城大殿即大安阁为基点，向北过皇城北门至外城东北门止，向南过宫城南门御天门至皇城南门明德门止。据考古勘察，这条中轴线宽达25米，沿线建筑秩序规整，宫城和皇城内的道路格局基本上以此中轴线对称分布。但元上都的外城建筑与道路分布较为自由，包括官署、街市、仓库、兵营和民宅等。外城北部为空旷的草地，应为草原族群的毡帐区域，这体现出元上都的营建既借鉴了中原地区都城规划理念，又与草原族群游牧习俗相融合。

元上都的都城营造过程，给元大都以及北京中轴线的建设提供了大量经验和借鉴。

The layout of Xanadu is square, with the outermost layer being the outer city. The southeast section of the outer city housed the imperial city, while the palace city was located slightly north of the center within the imperial city. A central north-south axis ran through both the palace and imperial cities. This axis was as wide as 25 meters, with orderly arranged buildings along its length. The layout of the roads within the palace and imperial cities was symmetrically distributed based on this central axis. However, the outer city's buildings and roads were more freely arranged, with the north part of the outer city being open grassland, likely used for tents by the nomadic groups, reflecting the fusion of the city's construction with the nomadic traditions of the steppe peoples.

一衣带水

Neighborly Bonds

中华文明辉映下的古代近邻都市

Ancient Cities of Close Neighbors
Impacted by Chinese Civilization

临水而建且规划严整的越南顺化古城
The ancient city of Hue in Vietnam, built by the water side and with a well-organized layout

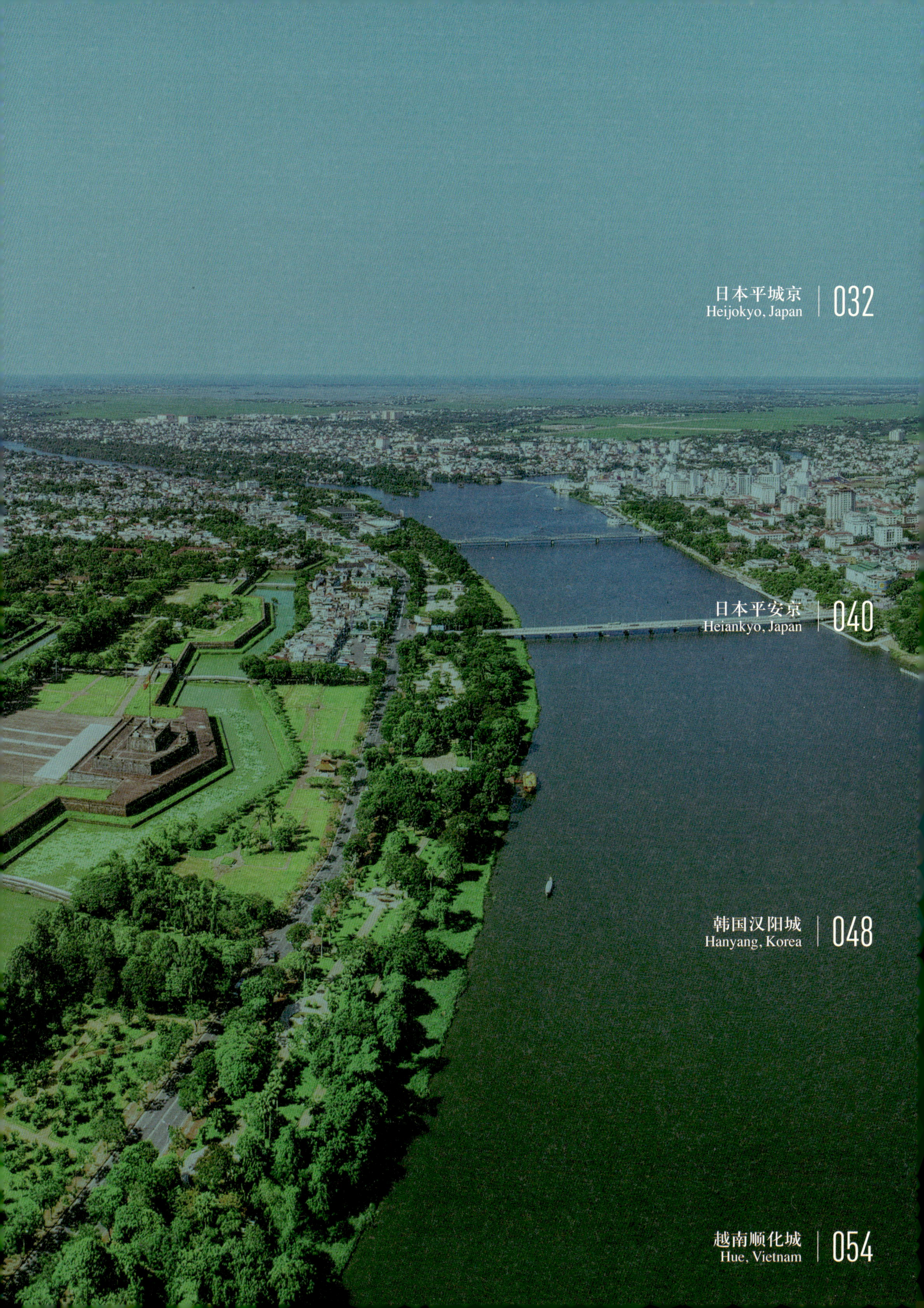

日本平城京的法隆寺，佛教刚传入日本时修建的寺院之一
Horyuji Temple in Heijokyo, Japan, one of the temples built when Buddhism was first introduced to Japan

日本平安京的东寺五重塔，古京都的象征
The Five-Story Pagoda of Toji Temple in Heiankyo, Japan, a symbol of ancient Kyoto

韩国汉阳城的昌德宫，被誉为"韩国的故宫"
Changdeokgung in Hanyang, Korea, known as the "Forbidden City of Korea"

越南顺化城，以四面围合的城墙和护城河拱卫重城
Hue in Vietnam, with its Imperial City encircled by walls and a protective moat on all four sides

　　中国在历史上曾经强烈地影响了周边国家和地区的文明与文化发展。作为古代都城建设特色的中轴线规划格局，也曾广泛地被一衣带水的日本、朝鲜半岛和越南等地区的古代政权所借鉴和吸收，用于当地都城的营建。

　　从 5 世纪到近代，周边地区持续不断地与中国进行深刻的文明与文化交流。以魏晋时期的邺城为代表的都城建造理念，就已被日本、朝鲜半岛等地所了解和模仿，比如日本的平城京、平安京的布局与曹魏邺城有明显的相似性。隋唐时期，东亚、东南亚区域的跨洋交流日益频繁，洛阳城等都城的堂皇气象不断地浸染周边地区，以《考工记》所载的诸如"前朝后市""左祖右社"等理想都市规划为范本的古代中国都城营建理念与实践，也在周边地区得到了一定程度的认同与推广。虽然这些都城的营建早于北京中轴线的出现，但追根溯源，它们与北京中轴线有着相似的久远渊源。及至明清时期，儒家风潮席卷周边日本列岛、朝鲜半岛及越南等地，在引领文化圈潮流的同时，也以其礼仪秩序、祖先祭祀、天地社稷等文化思想影响了周边地区的都城建设，儒家思想在一些都城规划与布局中得到体现。

　　周边地区在借鉴古代中国营城理念的同时，也因地制宜地进行取舍与扬弃。它们的都城建设或与本地山川地貌相适应，对来自古代中国的中轴线等思想有机使用甚至改弦更张；或与本土宗教和流行宗教相结合，在建筑布局和功能上体现出自身文化的特性，以特色鲜明的都市规划，满足自身的政治、文化、生活需求。

　　于是，当人们站在中国古代规划严整的都市，举目眺望周边的都市，会发现从 5 世纪至今，后者的都市轴线与北京中轴线相比，在规划理念、轴线与城市关系方面确实具有一定相似性，这种相似性来自中华文明与周边文明的交流互动，证明了中轴线思想的强大且持久的影响力；而后者在轴线规划格局、构成要素和景观形态方面，则与以北京中轴线为代表的古代中国都市轴线有显著不同，体现出营城文化上各美其美、百花齐放的形态。

Throughout history, China has significantly influenced the cultural development of neighboring countries and regions. The central axis planning, characteristic of ancient capital construction, was also widely adopted and assimilated by regions such as Japan, the Korean Peninsula, and Vietnam, where it was applied to the construction of their capitals.

From the 5th century to the modern era, these neighboring regions engaged in deep and continuous cultural exchanges with China. The concept of capital construction, exemplified by Ye City from the Wei and Jin periods, was understood and imitated in places like Japan and the Korean Peninsula. For example, the layouts of Japan's Heijokyo and Heiankyo show obvious similarities to that of Ye City during the Cao Wei period. During the Sui and Tang dynasties, transoceanic exchanges between East Asia and Southeast Asia became increasingly frequent, and the grandeur of capitals like Luoyang gradually influenced the surrounding regions. Chinese urban construction philosophy and practices, as outlined in texts like *Kaogongji*, which described ideal urban planning concepts such as "court in the front, market at the back" and "ancestral temples on the left, altars of land and grain on the right" were also recognized and promoted in neighboring regions. By the Ming and Qing dynasties, Confucianism swept across Japanese archipelago the Korean Peninsula, and Vietnam, and its cultural ideas, such as the order of rites, ancestor worship, and altars of land and grain, deeply influenced the construction of capitals in these regions.

While neighboring regions borrowed from ancient Chinese city-building concepts, they also made adaptations based on their own conditions. Their capitals were either adapted to local geographic features or integrated with indigenous or popular religions, reflecting their own cultural characteristics in terms of architectural layout and functionality, meeting their political, cultural, and everyday needs.

When comparing the urban central axes of neighboring regions with the central axis of Beijing, there are indeed certain similarities in planning philosophy and the relationship between the axis and the city, demonstrating the powerful and lasting influence of the central axis concept. However, there are also significant differences in terms of the layout, component elements, and visual form of these axes compared to the ancient Chinese urban central axes represented by Beijing Central Axis.

日本平城京
Heijokyo, Japan

日本奈良兴福寺及其五层宝塔
Kofukuji Temple in Nara, Japan and its five-story pagoda

日本平城京轴线示意图
Diagram of the axis of Heijokyo, Japan

7 世纪中期，日本开启影响深远的"大化改新"，打击豪族，推行律令，加强中央皇室的权威。694 年，皇室迁都藤原京，这是日本第一个条坊制的都城。此后到 710 年，日本元明天皇迁都平城京，此地作为奈良时代的都城，一直持续到 784 年桓武天皇迁都长冈京为止。

平城京位于今天日本奈良县郊区，其整体格局与隋唐长安城、洛阳城乃至更古老的曹魏邺城颇有相似之处，说明大化改新后，大量遣唐使从唐朝吸收了都城营建的理念，带回日本国内用于自己都城的建设。

平城京轮廓呈不太规则的正方形，东西宽 4 ~ 6 千米，南北长约 5 千米，采用条坊制布局，东西共 32 町，南北共 36 町，形成棋盘式规整的道路网络。城内另有宫城，位于都城北部中央高地上，俯视全城。宫城内的大极殿与宫城南面正门朱雀门相对，朱雀门向南的全城主街——朱雀大街直通都城正南门即罗城门，由此构成了城市中轴线。朱雀大街把全城分

割为东、西两部分，分别称为"左京""右京"。以朱雀大街为轴，两侧对称分布着东市、西市，以及药师寺、大安寺等重要寺院。左、右两京各自被划分为南北走向的 4 条坊带，每条坊带又被东西走向的路径所分割，构成棋盘状的町阵。

"古奈良历史遗迹"于 1998 年以遴选标准 (ii)、(iii)、(iv)、(vi) 成功列入《世界遗产名录》，遗产地包括平城宫迹和保留至今的东大寺、春日大社、兴福寺、药师寺、唐招提寺、元兴寺等 6 座古代庙宇及原始森林。该遗产展现了日本古代建筑与艺术演化发展的成就和文化上与中国的联系，但昔日的中轴线布局已然难寻。平城京遗址如今被奈良近郊城市所叠压，仅有部分遗迹可辨识。就轴线布局来说，平城宫朱雀门南还保留有朱雀大街局部历史遗迹，道路其他部分均已不存。与平城京中轴线相关的历史建筑仅剩下平城宫迹，后人在大极殿、朱雀门等重要建筑遗址上进行了展示性重建。

平城宫第二次大极殿遗址。所谓"第二次",指当时圣武天皇把都城迁回平城京后再次修建的大极殿
The Site of the Second Daigokuden at Heijo Palace. The term "second" refers to the Daigokuden (Great Hall of State) that was rebuilt when Emperor Shomu moved the capital back to Heijokyo and reconstructed the grand hall

平城宫遗址正门朱雀门 (复建)
The Suzaku Gate (reconstructed) at the Site of Heijo Palace

兴福寺中金堂 (复建)
The Central Golden Hall of Kofukuji Temple (reconstructed)

In 710, Japanese Empress Genmei moved the capital to Heijokyo, which remained the capital during the Nara period until 784. Heijokyo was located in what is now the outskirts of Nara Prefecture.

Its overall layout bore considerable resemblance to Chang'an and Luoyang from the Sui and Tang dynasties, as well as the even older Ye City of the Cao Wei period. The city had a roughly square shape, though not perfectly regular, with an east-west width of approximately 4 to 6 kilometers and a north-south length of about 5 kilometers. It adopted the Jo-Bo system (a grid-based urban layout), with 32 blocks from east to west and 36 blocks from north to south, forming an orderly grid of streets. Inside the city was the palace complex, situated on elevated ground in the center of

the north part of the capital, overlooking the entire city. The main hall of the palace (Daigokuden) faced the south main gate (Suzaku Gate), and Suzaku Avenue, the main thoroughfare running south from Suzaku Gate, led directly to the capital's south gate, forming the central axis of the city.

The "Historic Monuments of Ancient Nara" were inscribed on the *World Heritage List* in 1998. The heritage site includes the ruins of Heijo Palace as well as six ancient temples and a primeval forest that have been preserved to this day. This heritage site showcases the achievements in the evolution of ancient Japanese architecture and art, and the cultural ties with China, though the central axis layout of the past is now difficult to discern.

药师寺东、西三塔
The east and west pagodas of Yakushiji Temple

复建后的兴福寺中金堂被回廊柱础遗迹所环绕
The reconstructed Central Golden Hall of Kofukuji Temple is surrounded by the remains of the corridor's foundation stones

唐招提寺金堂，最初为东渡日本的鉴真和尚主持修建
The Main Hall of Toshodaiji Temple, originally built under the supervision of the Chinese monk Jianzhen, who traveled to Japan

奈良春日大社鸟居和石灯笼
The Torii and Stone Lanterns at Kasuga Taisha in Nara

元兴寺极乐坊本堂
The Main Hall of Gokurakubo at Gangoji Temple

中轴视角之日本古代都城
Ancient Japanese Capitals from the Central Axis Perspective

日本古代都城轴线最初的规划理念是为了象征皇权的神圣性，都城中轴线以居中道路纵贯宫城的朝堂正殿和全城，街巷格局以中轴线左右对称，这一点与北京中轴线的理念是相通的。但以平城京、平安京为代表的日本都城轴线并没有《考工记》这样的文献为指南，只是借鉴了中国魏晋至隋唐时代的都城格局，并无"左祖右社""面朝后市"的布局特征。而且由于营建较早，其轴线的构成要素也比较简单，祭奠祖先和祭祀神灵的礼仪建筑在城中并未形成严谨的对称关系，也没有在轴线上设置商业街市，整条轴线在景观上更接近唐长安的轴线形态。

The initial planning concept for the central axis of ancient Japanese capitals was intended to symbolize the sacred nature of imperial power. The central axis of the capital city was a road running through the palace's main hall and the entire city, with streets and alleys arranged symmetrically on either side of this axis. This is similar to the concept of Beijing Central Axis. However, unlike China, which had texts like *Kaogongji* as a guide, the layout of Japanese capitals such as Heijokyo and Heiankyo was based on the capital layouts from the Wei-Jin to Sui-Tang periods in China, and did not feature the "ancestral temple on the left, altar of land and grain on the right" or "court in the front, market in the back" configurations.

Furthermore, since these capitals were built relatively early, the elements of their axes were relatively simple. Ritual buildings for ancestor worship and deities did not follow a strict symmetrical arrangement within the city, nor was there a commercial street set along the central axis. The overall landscape of the axis was much more close to one in Chang'an in the Tang Dynasty.

东大寺，因建在平城京以东而得名，另有西大寺与之相对应
Todaiji Temple, named for being built to the east of Heijokyo, with Saidaiji (West Great Temple) as its counterpart

日本平安京

Heiankyo, Japan

鸟瞰京都，古代建筑群与现代建筑群交错
The Aerial view of Kyoto, where ancient and modern architectural complexes intertwine

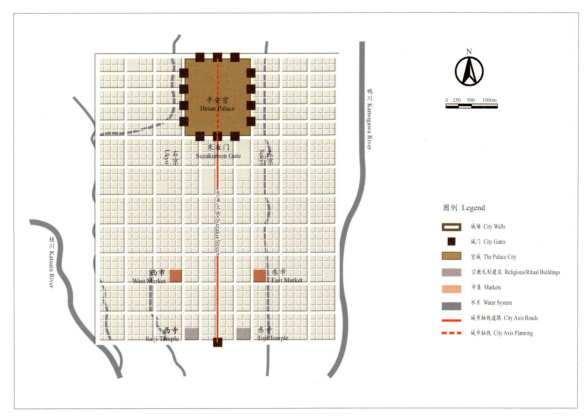

日本平安京轴线示意图
Diagram of the axis of Heiankyo, Japan

794年，日本桓武天皇将都城迁移到位于京都盆地北部的平安京，从此以后直到明治二年即1869年止，平安京一直是从中世纪到近代日本的都城，沿用千年。

与平城京一样，平安京也明显受到了中国古代都城规划理念的影响。整个都城为长方形，东西宽约4.5千米，南北长约5.2千米。城内设有宫城即"大内里"，位于都城北端中央，东西宽约1.2千米，南北长约1.4千米。

平安京有一条明显的中轴线，一条南北方向的朱雀大街从宫城的南门朱雀门向南延伸，将都城分成左、右两京，东市、西市和一些重要寺院分列中轴线两侧，对称分布。宫城之内中央偏东还设有"内里"，可以大体看作皇室的内廷，从布局上与唐长安大明宫有诸多相似之处，中央有南北方向的中轴线，排列了正殿紫宸殿和其他一些宫殿，是主要建筑所在，中轴线两旁左右对称分布一些次要殿宇和小院落。

古京都遗址（京都、宇治、大津城）于1994年以遴选标准(ii)、(iv)列入《世界遗产名录》，由包括二条城和诸多神社、寺院在内的17处历史遗迹构成，保存有日本平安时代传统风貌的寺院、园林、城堡等建筑群和遗址。该遗产的突出普遍价值主要体现在建筑和园林艺术层面而非城市布局方面，它见证了日本木结构建筑特别是宗教建筑的发展和日本园林艺术的变迁。经过千年岁月，平安京的棋盘式道路网络保持了下来，但是昔日的中轴线朱雀大街已然不见，很多对称布局的古代建筑也已不存，仅存东寺（又名教王护国寺）。

In 794, Emperor Kanmu moved the capital to Heiankyo, located in the northern part of the Kyoto Basin. From then until 1869, Heiankyo remained the capital of Japan, lasting through the medieval and early modern periods for over a millennium.

Like Heijokyo, Heiankyo was clearly influenced by the planning concepts of ancient Chinese capitals.

京都东寺金堂，该寺是平安京留存至今的唯一古建筑
The Main Hall of Tōji in Kyoto, the only ancient building from Heiankyo that still exists today

The city was rectangular, with an east-west width of approximately 4.5 kilometers and a north-south length of about 5.2 kilometers. The palace complex, known as "Daidairi", was located at the center of the north end of the city. Heiankyo had a prominent central axis, Suzaku Avenue, running north-south from the Suzaku Gate, extended southward and divided the capital into the Left and Right Capitals. The East Market, the West Market, and several important temples were symmetrically distributed on both sides of the central axis. The palace contained an inner precinct known as "Naidairi", located slightly east of the center, which can be roughly regarded as the imperial family's inner court. Its layout bore many similarities to the Daming Palace of Tang Chang'an.

The Historic Monuments of Ancient Kyoto were inscribed on the *World Heritage List* in 1994, consisting of 17 historical sites, including Nijo Castle and numerous shrines and temples. These monuments witness the development of Japanese wooden architecture, especially religious buildings, as well as the evolution of Japanese garden art. While the grid-like road network of Heiankyo has been preserved through the centuries, the central axis, Suzaku Avenue, has long disappeared, and many of the symmetrically arranged ancient buildings no longer exist.

京都西本愿寺
Nishi Honganji Temple in Kyoto

京都宇治喜撰桥和浮岛十三重石塔
Kisen Bridge and the Thirteen-Story Stone Pagoda on Floating Island in Uji, Kyoto

京都宇治平等院凤凰堂，以木造天盖、云中菩萨供奉像、凤凰金铜像、梵钟与壁画集于一堂而闻名于世
The Phoenix Hall of Byodoin in Uji, Kyoto, famous for its wooden canopy, cloud-dwelling bodhisattva statues, gilded bronze phoenix sculptures, temple bell, and murals

京都鹿苑寺舍利殿，有"金阁"的美名
The Reliquary Hall of Kinkakuji Temple in Kyoto, also known as the "Golden Pavilion"

京都慈照寺观音殿，有"银阁"之称，与"金阁"相对
The Kannonden of Ginkakuji Temple in Kyoto, known as the "Silver Pavilion", in contrast to the "Golden Pavilion"

京都高山寺石水院，以鸟兽人物戏画而闻名
The Sekisuiin at Kozanji Temple in Kyoto, famous for its "Animal Caricatures" scrolls

京都天龙寺本殿入口，该寺规模宏大，为"京都五山"（即天龙寺、相国寺、建仁寺、东福寺、万寿寺）之首

The entrance to the main hall of Tenryuji Temple in Kyoto, a large temple that is the foremost of the "Five Great Zen Temples of Kyoto" (Tenryu-ji, Shokoku-ji, Kennin-ji, Tofuku-ji, and Manju-ji)

京都龙安寺方丈庭园，也称石庭，是日本最有名的枯山水园林精品

The abbot's garden at Ryoanji Temple in Kyoto, also known as the "Rock Garden", is one of the most famous examples of Japanese Zen dry landscape gardening (Karesansui)

京都二条城本丸宫，该城为江户幕府开创者德川家康所筑，见证了江户幕府的兴起和衰亡

The Main Compound of Nijo Castle in Kyoto, built by Tokugawa Ieyasu, founder of the Edo Shogunate, which witnessed the rise and fall of the Edo period

京都清水寺本堂及悬空的舞台，该寺始建于8世纪，是京都最古老的寺院

The Main Hall and its hanging stage at Kiyomizu-dera in Kyoto, built in the 8th century, making it the oldest temple in Kyoto

韩国汉阳城
Hanyang, Korea

首尔昌庆宫，其中轴线为东西走向，正门弘化门和正殿明政殿均面向东方
Changdeokgung Palace in Seoul, featuring an east-west oriented central axis, with the main gate, Honghwamun, and the main hall, Myeongjeongjeon, both facing east

韩国汉阳城轴线示意图
Diagram of the axis of Hanyang, Korea

1392—1910 年，是朝鲜半岛的朝鲜王朝时代。1394 年，朝鲜王朝的都城——汉阳城建成，相比于半岛上早期的各种山城形制，这座都城虽然也建于绕城的山丘之上，却吸收了很多古代中国在平原上营建都城的理念。

由于受到山地地貌的影响，都城建筑在朝向方面并不统一，景福宫、太庙、昌德宫、文庙、成均馆坐北朝南，昌庆宫、庆熙宫及社稷坛则坐西朝东。宫城景福宫位于都城西北，宫墙北圆南方。

汉阳城内有南北和东西两个方向大道作为城市轴线。南北向轴线是景福宫正南门光华门向南大道，礼仪性较强，两侧分布着礼、枢、宪、兵、刑、吏六部官署。另一条东西向轴线则连接了都城东、西大门及庆熙宫、宗庙等重要建筑。城内依循《考工记》的规制，设立了"左祖右社"之布局，在景福宫东偏南建太庙，宫西偏南建社稷坛。

近代以来汉阳城多遭兵燹，城市格局遭到严重破坏，后恢复了部分建筑。其中保存较为完好的皇家宗庙和离宫昌德宫分别于 1995 年和 1997 年列入《世界遗产名录》，另存一些小段城墙和崇礼门、兴仁门等城门古建筑，被首尔的现代都市建筑所包围。

From 1392 to 1910, the Korean Peninsula was under the rule of the Joseon Dynasty. In 1394, Hanyang, the capital city of the Joseon Dynasty, was built. Unlike earlier mountain fortresses on the peninsula, this capital was also constructed on hills surrounding the city but incorporated many of the concepts used in ancient Chinese capitals built on plains. Due to the mountainous terrain, the orientations of the buildings in the capital were not uniform.

Within Hanyang, there were two main avenues serving as the city's axes: one running north-south and the other running east-west. The north-south axis, which extended south from the main gate of Gyeongbokgung, had a ceremonial function, with the six ministries located along both sides. The east-west axis connected the east and west gates of the capital, as well as important buildings such as Gyeonghuigung and Jongmyo Shrine(the royal ancestral shrine). Following the principles outlined in *Kaogongji*, the Jongmyo Shrine was built to the southeast of Gyeongbokgung, and the Sajik-Dong (the Altar of Land and Grain) was constructed to the southwest.

In modern times, Hanyang suffered significant destruction due to wars, and the city's layout was severely damaged, though some buildings were later restored. The well-preserved royal ancestral shrine, Jongmyo Shrine, and Changdeokgung were inscribed on the *World Heritage List* in 1995 and 1997, respectively.

首尔北汉山国家公园, 为汉阳城的北面屏障
Bukhansan National Park in Seoul, serving as the northern barrier for Hanyang

首尔景福宫正殿——勤政殿, 曾是朝鲜王朝举行重大庆典、重大外事活动的场所
The main hall of Gyeongbokgung in Seoul, known as the Geunjeongjeon, was the site where major celebrations and significant diplomatic events were held during the Joseon Dynasty

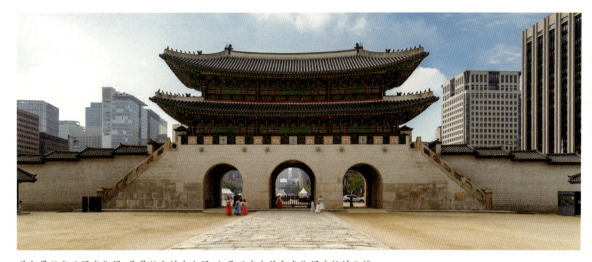

首尔景福宫正门光化门，是景福宫的南大门，也是世宗大路和光化门广场的北端
Gwanghwamun, the main gate of Gyeongbokgung in Seoul, is the south gate of the palace and the north end of Sejong-ro and Gwanghwamun Square

汉阳城城墙遗迹
The Site of the Hanyang City Walls

中轴视角之韩国汉阳城
Hanyang, Korea from the Central Axis Perspective

虽然汉阳城的规划受到了儒家文化的极大影响，宫城位于城内核心，轴线两侧设置宗庙、社稷坛等礼仪祭祀建筑，但其基础仍然是朝鲜半岛传统的山城防御布局，城郭依山势而建造，形态自由，以防御为要务。城中的南北轴线具有礼仪性功能，不过位置却并不在中央，也没有贯穿全城；城中的街巷和礼仪建筑的布局因地貌而显得松散自由，与北京中轴线布局的规整与统领全城形成鲜明对比。

Although the planning of Hanyang was greatly influenced by Confucian culture, with the palace complex at the core of the city and ceremonial buildings such as the Jongmyo Shrine and the Sajik-Dong set at the east and west sides of the central axis of Changdeokgung, its foundation still followed the traditional defensive layout of mountain fortresses on the Korean Peninsula. The city walls were built along the natural terrain, with a flexible structure prioritizing defense. The north-south axis within the city had a ceremonial function, but it was not centrally located, nor did it run through the entire city. The layout of streets and ceremonial buildings was loose and adapted to the terrain, in stark contrast to the orderly and centralized layout of Beijing Central Axis, which dominated the entire city.

首尔昌德宫仁政殿，昌德宫与昌庆宫因位于景福宫东边，因此合称"东阙"
Injeongjeon Hall at Changdeokgung in Seoul. Changdeokgung and Changgyeonggung are collectively called the "Eastern Palace" because they are located to the east of Gyeongbokgung

首尔庆熙宫崇政殿，该宫因位于景福宫西边，被称为"西阙"
Sungjeongjeon Hall at Gyeonghuigung in Seoul. This palace is called the "Western Palace" because it is located to the west of Gyeongbokgung

首尔昌庆宫明政殿
Myeongjeongjeon Hall at Changgyeonggung in Seoul

首尔景福宫正殿勤政殿，是古代朝鲜的国家最高级别建筑，殿前广场设有御道，两侧各有楼阁环卫，与北京故宫太和殿及殿前广
场布局相似
Geunjeongjeon Hall at Gyeongbokgung in Seoul, the highest-ranking building in ancient Korea. The square in front of the hall has a royal pathway, with pavilions on either side, similar to the layout of the Hall of Supreme Harmony and its square in the Forbidden City in Beijing

越南顺化城
Hue, Vietnam

越南顺化城午门外旗台，位于古城中轴线上
The flagpole outside the Meridian Gate of the Imperial City of Hue, Vietnam, located along the central axis of the ancient city

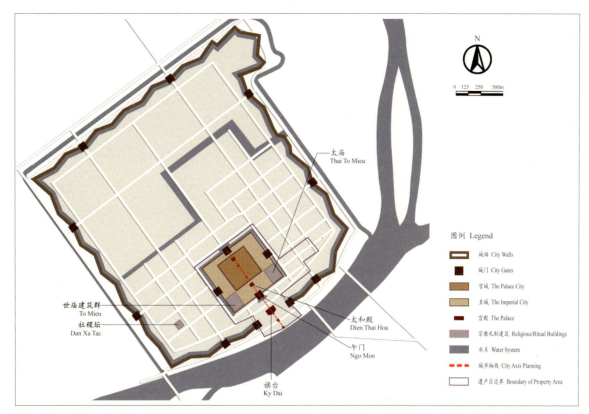

越南顺化城轴线示意图
Diagram of the axis of Hue, Vietnam

　　1802 年越南阮朝的嘉隆皇帝登基后，开始筹划在香江北岸建设都城——顺化城，经过近 30 年的营建终于大成。顺化城的规划受到古代中国都城营建理念的影响，并在防御城墙等建筑中融入了法国沃邦形式的建筑风格，形成东西合璧的形态。

　　整个都城大体呈方形，周长约万米，并适当顺应山川地貌影响而有所变化。都城内外由三重城墙所包围，由外到里依次是郭城、皇城和紫禁城（宫城）。设计者因地制宜地把皇城放置在都城的南部居中处，而宫城则位于皇城的核心。

　　宫城、皇城坐落于轴线之上并以此对称布局，郭城内道路格局也基本以轴线对称分布。顺化城的中轴线并非正南正北，而是偏东南至西北接近 45°。轴线从南城外安放火炮的旗台建筑群起始，向西北延伸，贯穿宫城中路核心建筑，到达皇城北门的北阙台。为了贴合"左祖右社"的布局理念，皇城的东南角设立太庙，与之相对的西南角设立世庙，而社稷坛则

设置于郭城内的西南角。

　　顺化历史建筑群于 1993 年以遴选标准 (iv) 列入《世界遗产名录》，包含顺化皇城和城内国子监、藏书楼等历史建筑或遗址，主要涉及顺化古城中轴线南段，即自迎凉亭至皇城北门。

　　In 1802, after Emperor Gia Long of the Nguyen Dynasty ascended the throne in Vietnam, he began planning the construction of the capital, Hue City, on the north bank of the Perfume River. After nearly 30 years of construction, it was finally completed. The planning of Hue City was influenced by the capital construction concepts of ancient China, while also incorporating elements of the French Vauban-style fortifications in its defensive walls and other structures, resulting in a blend of Eastern and Western architectural styles.

　　The city was generally square in shape, with a perimeter of about 10 kilometers, though it was adjusted to accommodate the natural terrain. The city was surrounded by three layers of walls, from the outermost to the innermost: the outer city, the imperial city, and the Forbidden City (the Palace City). The designers strategically placed the imperial city in the south central part of the capital, with the palace city at its core.

顺化城中轴线向南延伸，设有迎凉亭和敷文楼，位于旗台和香江之间
The central axis of Hue City extends southward, where the Nghinh Luong Dinh and the Phu Van Lau are located, between the Ky Dai (the Flag Tower) and the Perfume River

The palace city and imperial city were situated along the central axis and symmetrically laid out. The road network within the outer city was also symmetrically distributed based on this axis. However, the central axis of Hue City was not perfectly aligned north-south but was tilted at an angle of about 45 degrees from southeast to northwest. The axis began at the flag tower complex, where cannons were placed outside the southern city, and extended northwest through the core buildings of the palace city, ending at the north gate of the imperial city, the Bac Mon. To align with the layout concept of "ancestral temple on the left, altar of land and grain on the right", the Thai To Mieu (the Ancestral Temple) was built in the southeast corner of the imperial city, with the To Mieu (the Temple for Worshipping Emperors) placed in the southwest corner. The Dan Xa Tac (the Altar of Land and Grain) was set in the southwest corner of the outer city.

The Complex of Hue Monuments was inscribed on the *World Heritage List* in 1993. It includes the imperial city of Hue and historical buildings or sites such as the Quoc Tu Giam (National Academy) and the Royal Library. These primarily involve the south section of the central axis of the ancient city of Hue, extending from the Nghinh Luong Dinh to the north gate of the imperial city.

三层堆叠的旗台，上面设有火炮，曾有瞭望和防御外敌的功能
The Ky Dai is a three-tiered structure equipped with cannons, serving both as a lookout and for the defense against foreign enemies

顺化城午门，设有5个门洞，上有五凤楼
Hue City's Ngo Mon (Meridian Gate) features five gate openings, with the Ngu Phung Lau (the Five Phoenix Pavilion) situated on its top

顺化城太和殿是一座前后双体建筑，殿前设有牌坊和中道桥，为古代越南最高级别建筑，其形制与北京故宫太和殿类似，但规模较小
The Dien Thai Hoa (the Hall of Supreme Harmony) in Hue City is a dual-structure building with front and rear sections. In front of the hall, there is a ceremonial gate and the Central Path Bridge. It was the highest-ranking building in ancient Vietnam, similar in design to the Hall of Supreme Harmony in Beijing's Forbidden City, though smaller in scale

中轴视角之越南顺化城
Hue City, Vietnam from the Central Axis Perspective

顺化城的中轴线在很多方面与北京中轴线有相似之处，比如两者都体现出皇权至上和儒家礼仪的思想，在各自的城市结构中都处于核心位置，并贯穿多重城郭。但从规模上看，顺化城要小很多，因此城郭结构更为简单，轴线建筑群的尺度也较小。由于受到殖民时代的影响，顺化的建筑不仅有传统的木建筑样式，也融入了一些西洋建筑风格，与北京城的风貌有明显的差异。

The central axis of Hue City shares many similarities with the central axis of Beijing. Both reflect the supremacy of imperial power and Confucian ritual principles, and they occupy central positions in their respective city structures, running through multiple layers of city walls. However, in terms of scale, Hue City is much smaller, resulting in a simpler city wall structure and a more modest scale for the buildings along the axis. Due to the influence of the colonial era, Hue's architecture not only features traditional wooden structures but also incorporates some Western architectural elements, creating a noticeable contrast with the appearance of Beijing City.

顺化城西侧轴线，依次排列有显临阁、世庙、兴庙、奉先殿等建筑
Along the west axis of Hue City, buildings such as the Hien Lam Cac Pavilion, To Mieu, Hung Mieu, and Phung Tien Dien (Hall of Reverence) are arranged in sequence

顺化紫禁城后部回廊及建忠殿，该殿融合了欧洲建筑风格与越南传统建筑风格
The rear corridor of the Forbidden City and Dien Kien Trung in Hue City, which combines European architectural styles with traditional Vietnamese architecture

顺化城西部后宫区域的延寿宫
The Dien Tho Palace in the west rear Palace of Hue City

顺化城北墙正中的北阙台，台上修建有四方无事楼，为中轴线北端
The Bac Mon (Northern Gate) in the center of Hue City's north wall, with the Lau Tu Phuong Vo Su (Pavilion of No Worries in All Directions) built on its top, marking the north end of the central axis

顺化城护城河以及城门、城墙
The moat, gates, and walls of Hue City

南域宏光
Southern Splendor
东南亚与南亚的宗教、文化特色都市
Southeast and South Asian Cities Rich in Religious and Cultural Traditions

巴戎寺，位于吴哥通王城的中心
Bayon Temple, located at the center of Angkor Thom

吴哥窟，四周为护城河所环绕
Angkor Wat, surrounded by a moat

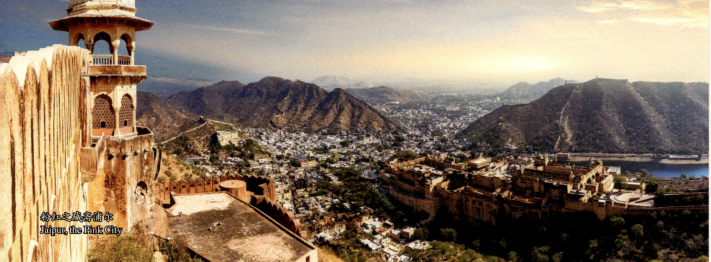

粉红之城斋浦尔
Jaipur, the Pink City

火山与大洋之间的日惹城市景观
The cityscape of Yogyakarta, situated between volcanoes and the ocean

新德里印度门
India Gate, New Delhi

作为世界文明古国，古代印度在思想、文化和宗教等方面强烈地影响了周边地区，特别对于南亚与东南亚地区来说，源自古代印度的佛教、印度教渗透社会的诸多方面，甚至在一些古代都城营建方面，也有宗教及其宇宙观的深远影响。

与古代中国《考工记》相似，古代印度也有一本描绘理想城市规划的文献《实利论》，该著作形成于公元前2—2世纪，以梵文写作，对城市选址、形态、护城河的形态尺寸、城墙位置尺寸、街道格局、神殿选址、王宫选址和朝向、其他诸方位的设施和居住区划分等方面均做出具体的规定。根据这一文献，理想的都城应为方形城郭，被厚重的城墙和护城河环绕。城内纵横各有三条大道，将城市划分为16个区域，城市中心应安置精神场所即神庙，代表王权的王宫应位于神庙以北；然后，从中心向外，不同的方位布置不同的功能区域，形成井然有序的城市布局。

7—16世纪，东南亚、南亚地区相继出现了一批强大的古代政权，比如吴哥王朝、素可泰王朝、阿瑜陀耶王朝、德里苏丹国、莫卧儿帝国等等。在古代印度营城思想的影响下，很多都城均以古印度文化理想都城范式为城市规划原型，比如高棉帝国都城吴哥通王城、印度拉贾斯坦邦的斋浦尔老城等。这些古代都市具有十字交叉的道路布局，呈现出一定的轴线特点。

近代，南亚、东南亚一些国家沦为西方列强殖民地，社会文化开始西风东渐。进入18世纪后，这些地区的城市和建筑发展又受到宗主国的城市规划理念与文化的影响，致使传统的城市布局发生转变，原本的十字布局与轴线布局发生改变，并被赋予了现代都市轴线的元素，比如印度尼西亚的日惹与印度新德里等城市，出现了三角轴线规划与几何图案的广场等近现代西方理念的实践。

相比于北京中轴线"以中为尊"和皇权至上的呈现风格，南亚和东南亚所展现的轴线更多与古代印度的宇宙观和西方近现代城市布局理念有强烈的关联性，对宗教、商贸乃至近现代行政制度与公众生活的考量更多。虽然同属亚洲东部的文明古国并在近现代受到西方冲击，但以北京中轴线为代表的东亚都市布局与南亚、东南亚城市布局双峰并立，分别为我们展现了古人不同的信仰体系和创造构思，展示出各自文化遗产的突出普遍价值。

Ancient India exerted a strong influence on surrounding regions in terms of thought, culture, and religion, and even in the construction of ancient capitals, religion and its cosmology had a profound impact. Similar to *Kaogongji* in ancient China, ancient India had a text describing the ideal city layout, *Arthashastra*, which was composed between the 2nd century BCE and the 2nd century CE. According to this text, the ideal capital should be a square framework, surrounded by thick walls and a moat. Inside, three main roads should run both north-south and east-west, dividing the city into 16 sections. The spiritual center, represented by a temple, should be placed in the middle of the city, while the palace, representing imperial authority, should be located to the north of the temple. Then, radiating outward from the center, different functional areas should be placed in specific directions, forming an orderly city layout.

From the 7th to 16th centuries, powerful ancient states emerged in Southeast Asia and South Asia. Under the influence of ancient Indian city-building concepts, many capitals were modeled on the ideal city paradigm of ancient Indian culture, featuring a grid layout with intersecting roads and displaying certain axial characteristics.

In modern times, the urban development and architecture of some South and Southeast Asian countries have been influenced by the urban planning concepts and cultures of their colonial powers, leading to transformations in traditional city layouts. The original grid and axial layouts were altered and infused with elements of modern urban axes, such as the triangular axis layout and geometric patterns in squares, reflecting Western ideas in cities like Yogyakarta in Indonesia and New Delhi in India.

Compared to the central axis of Beijing, which emphasizes "the centrality and the supremacy of imperial power", the axes displayed in South and Southeast Asia are more closely related to ancient Indian cosmology and Western modern urban planning concepts. These cities reflect greater consideration for religion, commerce, and even modern administrative systems and public life.

柬埔寨吴哥窟

Angkor Wat, Cambodia

大乘佛教寺院巴戎寺，以微笑四面佛雕像闻名于世
Bayon Temple, a temple of Mahayana Buddhism, is world-renowned for its smiling four-faced Buddha statues

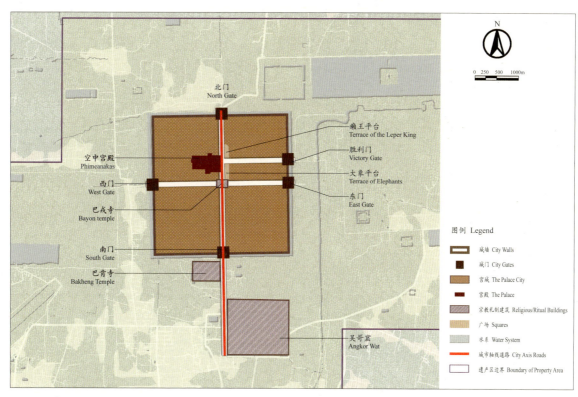

北门
North Gate

癞王平台
Terrace of the Leper King

胜利门
Victory Gate

空中宫殿
Phimeanakas

大象平台
Terrace of Elephants

西门
West Gate

东门
East Gate

巴戎寺
Bayon temple

南门
South Gate

巴肯寺
Bakheng Temple

吴哥窟
Angkor Wat

图例 Legend

城墙 City Walls
城门 City Gates
宫城 The Palace City
宫殿 The Palace
宗教礼制建筑 Religious/Ritual Buildings
广场 Squares
水系 Water System
城市轴线道路 City Axis Roads
遗产区边界 Boundary of Property Area

吴哥通王城轴线示意图
Diagram of the axis of Angkor Thom

9世纪初的中南半岛上，阇耶跋摩二世统一高棉，建立吴哥王朝。约50年后，耶输跋摩一世迁都并兴建了吴哥城。至15世纪，暹罗人攻入吴哥城，高棉人被迫迁都金边，彻底放弃了吴哥，任由蔓草杂林吞噬了庙宇。

吴哥窟是9—15世纪吴哥王朝都城的辉煌遗迹，于1992年以遴选标准(i)、(ii)、(iii)、(iv)列入《世界遗产名录》。吴哥的世界遗产由三个部分组成，主要部分是以大吴哥通王城和吴哥寺、巴方寺、塔布隆寺、茶胶寺等几十处吴哥时期的庙宇组成的遗址群，另外还包括位于吴哥城北的女王宫，以及位于南侧的巴空寺等十几处寺庙遗址。整个吴哥窟建筑群方圆约20平方千米。

吴哥通王城为方形，四周筑有城墙，墙外设有护城河。四面城墙中间设置城门，城内四方道路构成十字形，直通城市中心。城中心的巴戎寺向南到吴哥寺之间的道路，形成了城市的南北轴线。其中巴戎寺是城中最大的神庙，代表神的居所。巴戎寺西北坐落着皇家宫殿，南北轴线从巴戎寺向北联络宫殿，并延伸到城北门；向南延伸到城南门并通往城外的吴哥寺、巴肯寺等大型寺院。

可以看出，通王城的城市规划与建筑是古代印度的宇宙观和理想都市构想的实践。

Angkor Wat is the magnificent remnant of the capital of the Angkor Empire from the 9th to the 15th century, which was inscribed on the *World Heritage List* in 1992. The World Heritage Site of Angkor is composed of three parts. The main section consists of a complex of temple ruins from the Angkor period, including Angkor Thom, Angkor Wat, Baphuon, Ta Prohm, and Chau Say Tevoda, among dozens of other temples. It also includes the Banteay Srei temple located north of Angkor, as well as several temple ruins to the south, such as Bakong. The entire Angkor complex spans an area of about 20 square kilometers.

Angkor Thom is square in shape, with walls built on all sides and a moat surrounding them. Gates are set in the middle of each wall, and roads within the city form a cross-shape, leading directly to the center of the city. The road from Bayon Temple at the city's center, extending south to Angkor Wat, forms the city's north-

吴哥通王城城门及道路两旁排列的雕像
The gate of Angkor Thom and the statues lined along the sides of the road

空中宫殿，位于巴戎寺西北
The Phimeanakas (Heavenly Palace), located northwest of Bayon Temple

巴方寺一角
A corner of Baphuon Temple

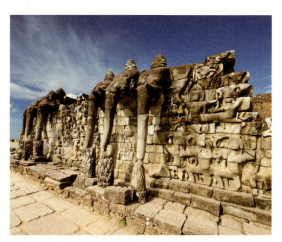

吴哥通王城的斗象台
The Terrace of the Elephants at Angkor Thom

south axis. Among them, Bayon Temple is the largest temple in the city, representing the dwelling place of the gods. Northwest of Bayon Temple lies the Royal Palace, with the north-south axis extending from Bayon Temple northward to connect with the palace and further to the north gate of the city. To the south, the axis extends to the south gate and leads outside the city to Angkor Wat, Phnom Bakheng, and other large temples. This city planning and architecture of Angkor Thom can be seen as a reflection of ancient Indian cosmology and the ideal city concept.

The architecture of Bayon Temple reflects the concept of Mount Meru in the Buddhist worldview

巴戎寺内部长廊
The inner corridor of Bayon Temple

中轴视角之柬埔寨吴哥窟
Angkor Wat, Cambodia from the Central Axis Perspective

　　吴哥窟的都市轴线受到了古印度文化的影响，城市规划体现了其独特的宇宙观。根据曼达拉（Mandala）模型，在人类居住的圆形大陆的中央耸立着世界中心须弥山，围绕须弥山由守护世界八个方位的八大守护神排列形成圆形区域，因此吴哥窟具有强烈的宗教象征意义，与北京中轴线所代表的儒家礼仪秩序和天人合一的宇宙观形成巨大的反差。

　　The urban axis of Angkor Wat was influenced by ancient Indian culture, and the city's layout reflects its unique cosmology. According to the Mandala model, Mount Meru, the center of the world, rises in the middle of a circular continent where humans reside. Mount Meru is surrounded by eight gods guarding the eight directions of the world, forming a circular region. As a result, Angkor Wat holds strong religious symbolic meaning, creating a stark contrast with the central axis of Beijing which represents Confucian ritual order and the cosmology of harmony between heaven and humanity.

巴戎寺建筑体现了佛教世界观中须弥山的观念
The architecture of Bayon Temple reflects the concept of Mount Meru in the Buddhist worldview

塔布隆寺，位于吴哥通王城东1千米
Ta Prohm Temple, located 1 kilometer east of Angkor Thom

圣剑寺
Preah Khan Temple

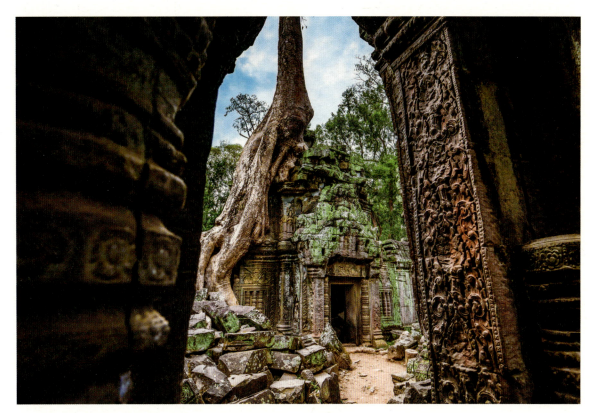

塔布隆寺，因古树缠绕古迹而形成独特的景观
Ta Prohm Temple, known for its unique landscape where ancient trees intertwine with the ruins

女王宫，建筑材料为红色砂岩，位于吴哥通王城西北方
Banteay Srei, constructed from red sandstone, located northwest of Angkor Thom

印度斋浦尔

Jaipur, India

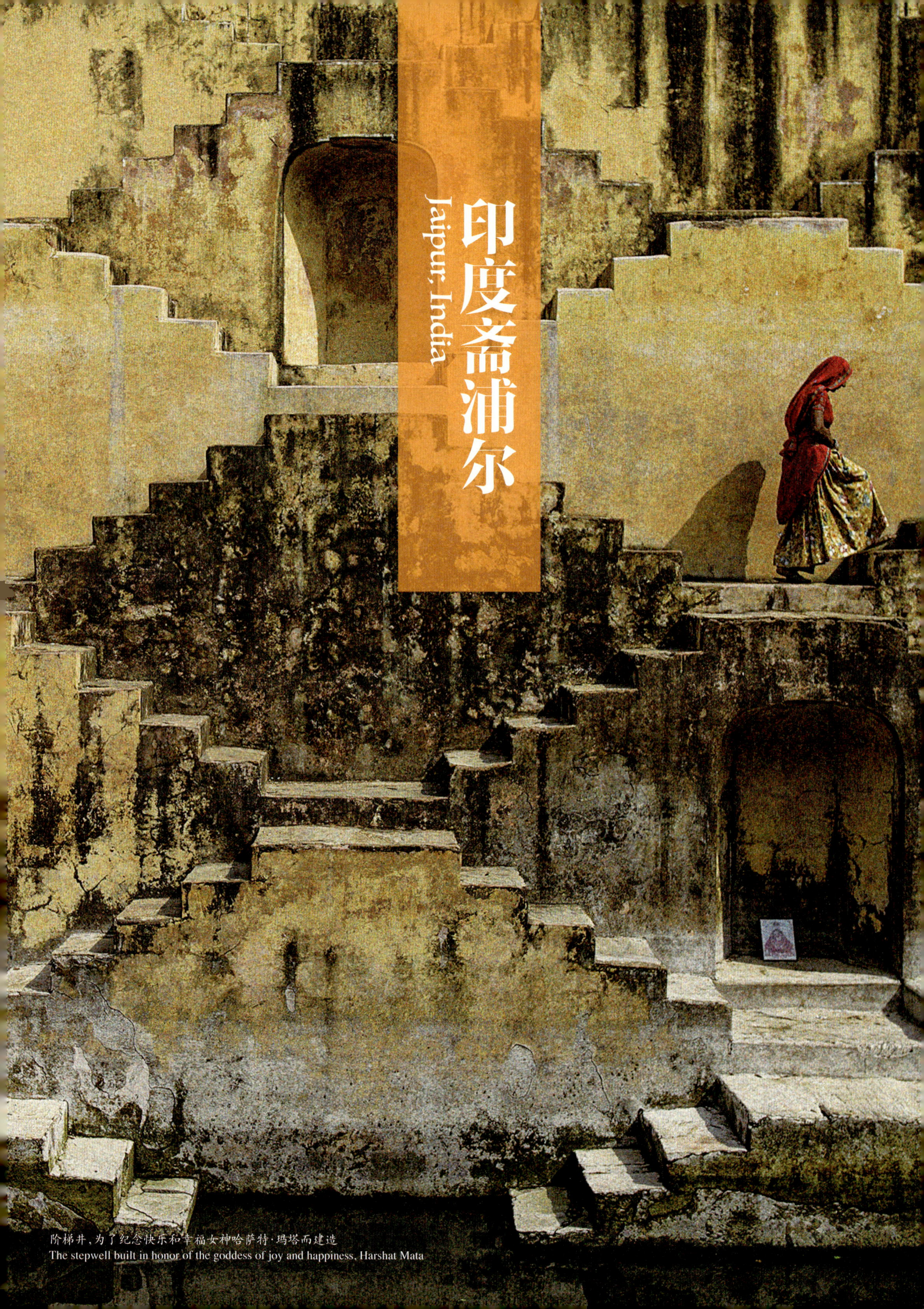

阶梯井，为了纪念快乐和幸福女神哈萨特·玛塔而建造

The stepwell built in honor of the goddess of joy and happiness, Harshat Mata

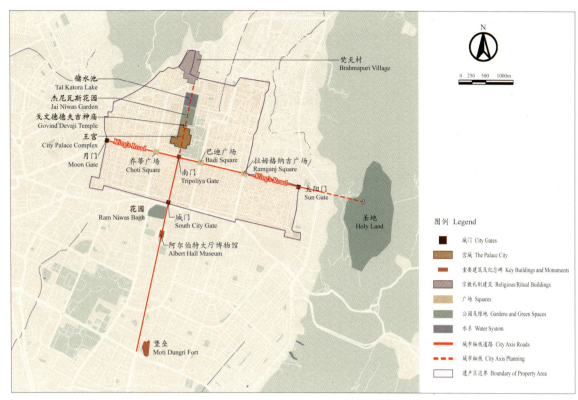

斋浦尔城市轴线示意图
Diagram of the urban axes of Jaipur, India

18世纪20年代，印度拉贾斯坦邦的统治者为了解决人口增长和缺水问题，兴建新的都城——斋浦尔。今天的斋浦尔老城建设于18世纪，而新城则是19世纪在老城基础上向外特别是向南扩张形成的。斋浦尔城于2019年以遴选标准(ii)、(iv)、(vi)成功列入《世界遗产名录》，遗产区包括由城墙围绕的斋浦尔老城，老城段的城市轴线也在其中。

斋浦尔老城的格局以800米×800米的正方形模块为基础，9个模块构成"井"字形的网格状空间布局，但由于西北角有山阻碍，因此建设者把西北模块移动到了东南角。在斋浦尔老城的9个街区中，中央和正北街区属于政权管理建筑和宫殿，包含神庙；另外7个街区属于民众生活区域，且在网格交点街角布置了广场和市集，以满足贸易和商业、手工业发展。

老城区有东西、南北两条十字交叉的轴线，东西向为主轴线，从王宫南侧横向展开，从西侧的城门一直通向东侧的山丘圣地；南北向轴线从南侧城门开端，经王宫正南门穿行向北，联络了宫殿、庭院、储水池及神庙。19世纪开始，南北向轴线随城市扩张而向南延伸，直至一处山丘止。

整个老城把古代印度理想都市布局、莫卧儿王朝时期建筑风格乃至西方城市规划理念融合在一起，并因为沿街建筑被漆成粉红色而有着"粉红城市"的美誉。

In the 1720s, the ruler of Rajasthan, India, began constructing a new capital, Jaipur, to address population growth and water shortages. Today's old city of Jaipur was built in the 18th century, while the new city expanded outward, particularly to the south, in the 19th century based on the old city. In 2019, Jaipur was inscribed on the *World Heritage List*, with the heritage area including the walled old city, along with the city axis within the old city.

The layout of Jaipur's old city is based on square modules measuring 800 meters by 800 meters, with nine modules forming a grid-like structure. However, because a mountain obstructed the northwest corner, the builders shifted the northwest module to the southeast

乔浦尔色彩艳丽的城市宫殿
The vibrant City Palace of Jaipur

corner. Of the nine blocks in Jaipur's old city, the central and north blocks were designated for administrative buildings and palaces, including temples, while the other seven blocks were residential areas. At the intersections of the grid, squares and markets were set up to support trade, commerce, and craftsmanship.

The old city has two intersecting axes: the main east-west axis, which extends laterally from the south side of the palace, running from the west city gate to the sacred hills in the east; and the north-south axis, which begins at the south city gate and passes through the south gate of the palace, connecting the palace, courtyards, water reservoirs, and temples. Starting in the 19th century, the north-south axis extended further south as the city expanded, reaching a hill at its south terminus.

The entire old city integrates ancient Indian ideal city planning, Mughal architectural style, and even Western urban planning concepts. Because the buildings along the streets are painted pink, Jaipur is affectionately known as the "Pink City".

斋浦尔城市宫殿
The City Palace of Jaipur

曼萨加尔湖中的水宫，位于斋浦尔老城轴线的向北延长线上
The Jal Mahal (Water Palace) in the Man Sagar Lake, located on the north extension of the old city's axis in Jaipur

印度文化中的"同心城圈"模型
The "Concentric Circles" Model in Indian Culture

印度文化影响下的都城以"中央神域"向周边城区呈现等级差异，距离中心越近的位置最佳，构成"同心城圈"的基本模型。内层城圈有王宫和"四姓共居的最佳居住地"，代表国王和服务于神域的四姓居住的区域，中间城圈设有官厅、官库等服务王权的设施，外圈是商人、手艺人的居所和市场等。因此，都城的基本格局具有很强的中心性，轴线的方向性或指向性并不明显。

城市宫殿中印度风格的建筑装饰
The Indian-style architectural decorations in the City Palace

Under the influence of Indian culture, the layout of capitals presents hierarchical differences radiating from a "central sacred domain" to the surrounding urban areas. The closer a location is to the center, the more desirable it is, forming the basic model of "concentric circle". The inner circle contains the imperial palace and the "ideal residences for the four social classes", representing the area where the king and the four classes serving the sacred domain reside. The middle circle includes facilities such as government offices and warehouses that serve the imperial power, while the outer circle is home to merchants, artisans, and markets. Thus, the basic layout of the capital exhibits a strong centrality, with the axis's directionality or orientation being less pronounced.

简塔曼塔天文台，位于城市宫殿东南侧
Jantar Mantar Observatory, located southeast of the City Palace

琥珀堡，位于水宫继续向北的延长线上
Amber Fort, located further north along the extension of the axis beyond the Water Palace

印度尼西亚日惹

Yogyakarta, Indonesia

默拉皮火山俯视下的日惹
The Overlook of Yogyakarta and Mount Merapi

日惹城市轴线示意图
Diagram of the urban axis of Yogyakarta

作为爪哇岛上曾经的日惹苏丹国的首都，日惹历史城区兴建于18世纪，它有一条明显的近南北向轴线，延伸达6千米，北端起自默拉皮火山，南端到达南海（印度洋的一部分）之滨，此火山与大海象征着爪哇文化宇宙观中的守护神和南海女王的居所。

"日惹的宇宙轴线及其历史地标"于2023年以遴选标准(ii)、(iii)列入《世界遗产名录》。在这一遗产地，纪念碑、宫殿及其城门、广场以及内部建筑群、大清真寺建筑群、行政建筑群在南北方向的城市轴线上分布，宫殿位居轴线中部，强调王权的威严；轴线连接火山与海洋，象征着天神、人类与自然之间具有和谐的关系，若干建筑组合还象征着整个人类从出生、婚姻到死亡的生命周期。除了建筑与布局等物质遗产外，日惹的宇宙轴线上还承载着祭拜仪式、加冕典礼和节日等非遗活动，体现了爪哇文化关于人类生命的哲学思想。

进入20世纪，日惹的宇宙轴线又经过西方规划师的设计与建设，日惹成为借助城市规划展现东南亚地区爪哇文化中宇宙学和哲学思想的重要实例。

As the capital of the former Sultanate of Yogyakarta on Java Island, the historic city of Yogyakarta was established in the 18th century. It features a prominent north-south axis that extends for 6 kilometers, starting from Mount Merapi in the north and reaching the south coast by the Indian Ocean. This volcano and the sea symbolize the patron gods and the residence of the Queen of the South Sea within the Javanese cosmological worldview.

The "Cosmic Axis of Yogyakarta and Its Historical Landmarks" was inscribed on the *World Heritage List* in 2023. Within this heritage site, monuments, palaces, city gates, squares, internal architectural groups, large mosque complexes, and administrative buildings are distributed along the north-south urban axis, with the palace situated in the middle to emphasize imperial authority. The axis connects the volcano and the ocean, symbolizing the harmonious relationship between deities, humanity, and nature. Several architectural combinations also represent the human lifecycle, from birth to marriage and death. In addition to tangible

日惹王宫及周边城市景观
Yogyakarta Palace and surrounding urban landscape

日惹南广场
The Southern Square of Yogyakarta

heritage such as buildings and layouts, Yogyakarta's cosmic axis encompasses intangible cultural heritage, including rituals, coronations, and festivals, reflecting the philosophical thoughts of Javanese culture regarding human existence.

In the 20th century, the cosmic axis of Yogyakarta underwent design and construction by Western planners, making Yogyakarta a significant example of how urban planning can express cosmology and philosophical thought within Javanese culture in Southeast Asia.

时常喷发的默拉皮火山强烈地影响了日惹地区的宇宙观
The frequently erupting Mount Merapi strongly influences the worldview of the Yogyakarta region

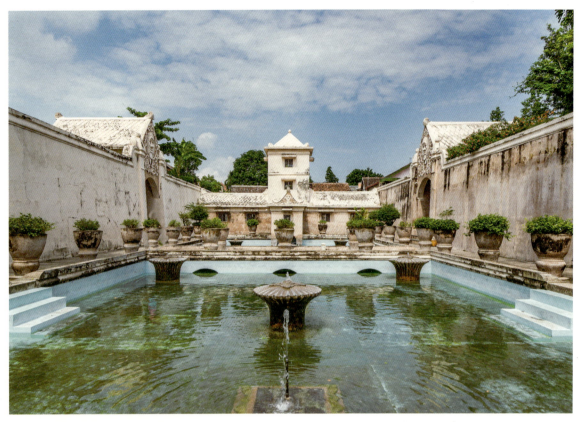

位于日惹中轴线上的塔曼萨里水城堡
Taman Sari Water Castle, situated on the Yogyakarta central axis

日惹弗登堡博物馆
Fort Vredeburg Museum in Yogyakarta

日惹纪念碑
Tugu Monument

大清真寺
Great Mosque

印度新德里
New Delhi, India

新德里旧国会大厦
New Delhi's Old Parliament House

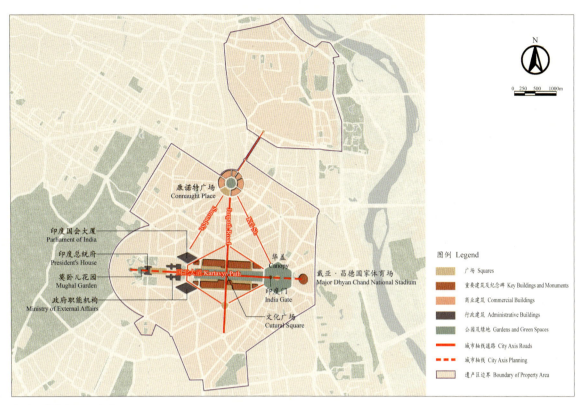

新德里城市轴线示意图
Diagram of the urban axes of New Delhi

作为英国殖民者选定的印度新首都，新德里奠基于1911年，在第一次世界大战后开始加快建设，并于1931年竣工。

在多中心城市模式理念的指导下，新德里城区内的总统府、印度门和康诺特广场构成三角形轴线结构：主轴线即东西方向连接总统府和印度门的国王大道（今责任大道），总统府兼具西方和印度建筑风格，是轴线的西端；向东沿着国王大道两侧，分布着议会、法院等行政机构，以及林荫道和绿化带，博物馆、艺术馆等文化建筑分列两旁，直通到东端六边形的广场绿地，绿地上矗立着纪念性建筑——印度门。三角形轴线另一端点康诺特广场位于北面，从广场出发，既有南北轴线与国王大道垂直相交，又有两个"斜边"通向总统府和印度门。康诺特广场为圆形喷泉广场，并建有内外环形建筑及拱廊，属于新德里的商业中心。

作为20世纪早期现代主义城市规划的实例，新德里的营建受到了当时田园城市思想的影响，并与当地印度宗教、文化和气候条件相结合，在建筑上形成了欧亚合璧的风格。

As the new capital of India selected by British colonizers, New Delhi was founded in 1911, with its construction accelerating after World War I and completed in 1931.

Guided by a multi-centered urban model, the city layout features a triangular axis structure connecting President's House, India Gate, and Connaught Place: the main axis runs east-west along Rajpath (now known as Kartavya Path), linking President's House, an architectural blend of Western and Indian styles, at the west end, with administrative buildings like Parliament and courts, shaded boulevards, and cultural institutions lining on both sides, leading to the hexagonal green space at the east end, where India Gate stands. Connaught Place, located to the north, serves as another endpoint of the triangular axis, intersecting with the north-south axis and featuring "diagonal" routes leading

新德里印度门
India Gate in New Delhi

to President's House and India Gate. This circular fountain square, surrounded by inner and outer arcades, serves as New Delhi's commercial hub.

As an example of early 20th century modernist urban planning, New Delhi's construction was influenced by the garden city movement, integrating local Indian religious, cultural, and climatic conditions to create a unique Eurasian architectural style.

印度总统府
President's House

圆形康诺特广场区域一角
A corner of the circular Connaught Place area

中轴视角之印度新德里
New Delhi, India from the Central Axis Perspective

　　作为20世纪兴起的都市，新德里的城市轴线颇具现代性，它以三角形为原型的放射状形态，成为多中心、开放性城市结构的一部分，引导城市发展。城市轴线由象征现代行政制度和公共生活的政府办公建筑、纪念性建筑、广场和绿化景观构成。而北京中轴线以明清宫城和现代纪念性建筑为核心，串联起大量礼仪祭祀建筑和城市管理设施，使得北京与新德里的城市风貌有显著的差异。

　　As a city that emerged in the 20th century, the urban axis of New Delhi embodies modernity. Its radial form, based on a triangular prototype, becomes part of a multi-centered, open urban structure that guides urban development. The urban axis consists of government buildings symbolizing modern administrative systems and public life, along with monumental structures, squares, and green landscapes. In contrast, Beijing Central Axis, centered on the Ming and Qing palace complexes and modern monumental buildings, connects a multitude of ritual and ceremonial structures as well as urban management facilities, resulting in a significant difference in the urban character between Beijing and New Delhi.

向西通往总统府的中轴线大街
The central axis street leading west to the President's House

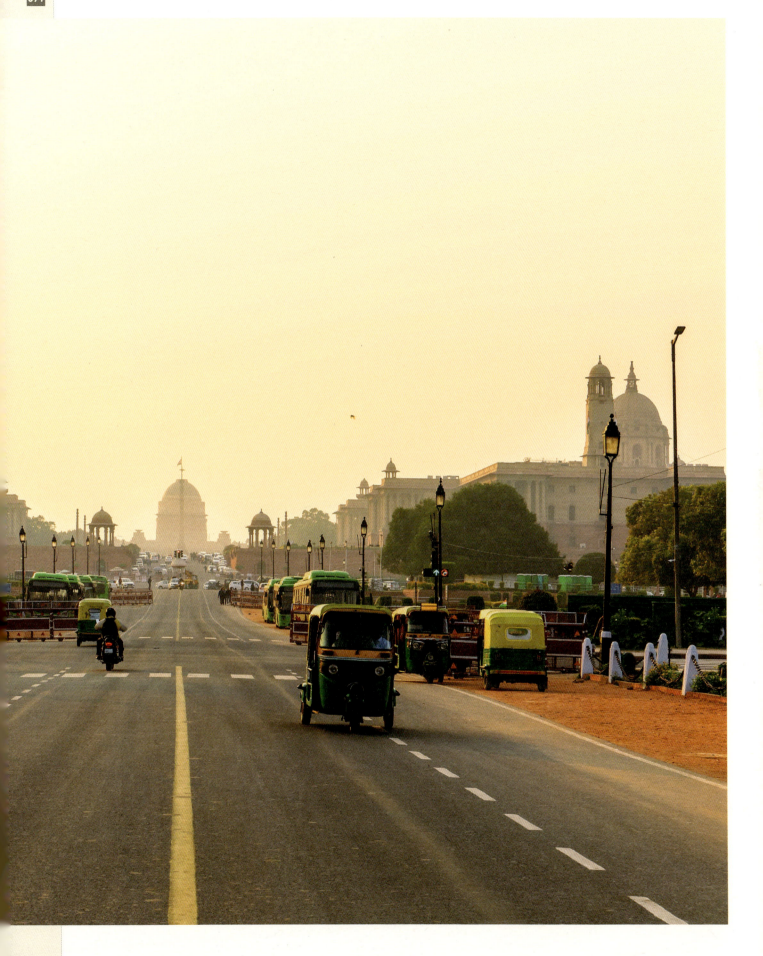

胡马雍陵，位于从印度门向外东南辐射的道路旁

Humayun's Tomb, located along the road radiating southeast from the India Gate

德里红堡，位于印度门东侧垂直中轴线的东西方向街道的北方
Red Fort situated to the north of the east-west street along the vertical central axis of the India Gate

康诺特广场，新德里三角形轴线的一个端点
Connaught Place, one endpoint of the triangular axis in New Delhi

贾玛清真寺，位于印度门北方，紧邻德里红堡
Jama Masjid, located north of the India Gate, adjacent to the Red Fort

殊方壹皇

Majestic Cities Afar

规划严整的欧洲著名都市
Europe's Prominent Cities with
Immaculate Urban Planning

罗马历史城区
Historic Center of Rome

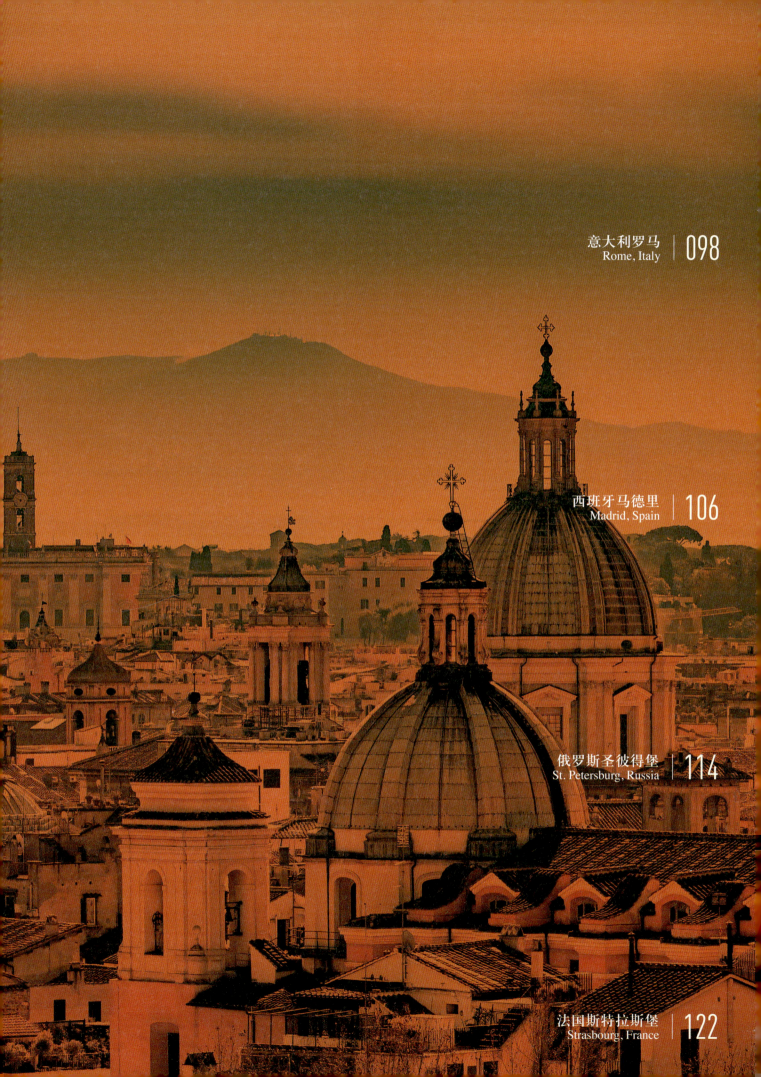

罗马
Rome

马德里
Madrid

圣彼得堡
St. Petersburg

斯特拉斯堡
Strasbourg

在宗教色彩浓郁的古代欧洲，城市轴线的出现相当久远。在古希腊甚至更早的时期，轴线设计就已在建筑与空间设计中有所应用。比如在雅典卫城，作为朝圣线路的雅典娜节日大道成为城市空间的组织基线，联系了城市里各组建筑群，对城市的发展起到了引导作用。

到了古罗马时期，人们热衷于设计城市广场和大型公共建筑，广场开始以轴线组织空间，以方正的平面、连续的柱廊、宏大的拱券塑造公共空间，比如罗马的帝王广场遗址群就是实例之一。而在做军事用途的营寨城池中，十字交叉或"T"字交叉的干道成为方形城池内的道路网络主干，突出其防御功能。

文艺复兴之后，人文主义和理性主义思想为欧洲城市布局注入新的活力。特别是16世纪巴洛克思潮兴起，城市轴线的规划方式在城市建设中得到了广泛应用，著名案例就是罗马城的改造，一系列方尖碑设置在城市格局的既有空间中，建构出影响城市发展的轴线系统，通过街道构建起交通和视觉联系，使罗马的城市空间凝集为一个整体，同时又因轴线系统的引导，使城市空间具有开放性。

进入17世纪后半叶，巴洛克艺术思想又被用于园林景观设计之中，而这样的园林景观又广泛用于城市与宫廷营建，法国的凡尔赛宫、沃·勒·维贡特花园的景观皆体现了轴线对称的、纯粹的、完美的几何构图。一个世纪之后，当人们对巴黎进行改造时，古典主义为指导的城市轴线以林荫大道和沿街立面风格统一的建筑形式，塑造出统一的轴线景观，广场、纪念碑、雕塑、植物景观成为轴线的重要节点，通过放射形道路做轴线，把城市各个区域联系起来，打造出协调统一的城市空间。

巴洛克风格和古典主义城市轴线的规划理念风靡欧洲，影响到了新建或改造的都市，如马德里、圣彼得堡、斯特拉斯堡等地的城市建设，甚至影响到了美洲新大陆的一些城市建设。

由于文化底蕴与发展脉络的差异，这些欧洲著名都市的轴线系统与北京中轴线有着明显的差异。受文艺复兴影响，以及应近代城市功能的要求，前者更加凸显宗教精神、文艺气息，更强调轴线系统的几何美感；而北京中轴线更强调儒家礼制秩序和天人合一等哲学理想。两类轴线系统的都市虽相隔千山万水，却又相映生辉。

In the religiously vibrant ancient Europe, the emergence of urban axes dates back quite far. As early as ancient Greece and even earlier, axial design was applied in architecture and spatial design. For instance, in the Acropolis of Athens, the processional avenue for the Athena festival became the organizing baseline of urban space, linking various architectural groups and guiding the city's development.

During the time of ancient Rome, there was a keen interest in designing urban squares and large public buildings. Squares began to be organized along axes, shaping public spaces with rectangular layouts, continuous colonnades, and grand arches, as exemplified by the ruins of the Imperial Forum in Rome.

After the Renaissance, humanism and rationalism infused new vitality into European urban layouts. The planning approach of urban axes became widely applied in city construction. A notable case is the transformation of the city of Rome, where a series of obelisks were integrated into the existing urban layout, creating an influential axis system that linked streets for traffic and visual connection, thus coalescing Rome's urban space into a cohesive whole while maintaining an open character.

In the 18th century, when Paris underwent transformation, classical principles guided the urban axis to shape a unified landscape along tree-lined avenues and cohesive building facades. Squares, monuments, sculptures, and plant landscapes became key nodes along the axes, connecting various city areas through radial roads and creating a harmonious urban space.

The planning concepts of Baroque and classical urban axes became fashionable in Europe, influencing the construction of newly built or renovated cities such as Madrid, St. Petersburg, and Strasbourg, and even impacting some cities in the New World.

Influenced by the Renaissance and in response to modern urban functional requirements, the former emphasized religious spirit and artistic atmosphere, highlighting the geometric beauty of axis systems; whereas Beijing Central Axis emphasizes Confucian ceremonial order and the philosophical thought of harmony between heaven and humanity.

意大利罗马
Rome, Italy

圣彼得大教堂内部
Interior of St. Peter's Cathedral

罗马城市轴线示意图
Diagram of the urban axes of Rome

最初的罗马城于公元前753年在台伯河畔建立，成为古罗马数百年的都城和当时地中海文明的政治、文化中心。从4世纪开始，这里成为基督教世界的精神中心，因此有"基督教罗马"之称。16世纪，为了串联7座重要的教堂，为前来朝拜的教众提供道路和方向，教皇西克斯图斯五世对罗马城进行了改造，使得罗马城形成了由一系列发散状道路构成的轴线系统，最知名的案例是波波洛广场，三条道路从此地出发，通往全城的重要教堂、广场以及古罗马遗迹如斗兽场等。城市道路的重要节点上竖立了纪念性方尖碑和若干喷泉，给公众提供方位信息以及视觉美感。这些改造被欧洲建筑学界视为巴洛克风格城市规划实践的开端。

"罗马历史中心"于1980年以遴选标准(i)、(ii)、(iii)、(iv)、(vi)列入《世界遗产名录》，包括17世纪由城墙围起的城区以及城墙外的圣保罗大教堂。这一世界文化遗产展现了古典

时期、文艺复兴和巴洛克时期等多个历史时期的城市发展成就，特别是包含了形成于巴洛克时期的城市轴线。

The original city of Rome was founded in 753 BCE on the banks of the Tiber River, serving as the capital of ancient Rome for centuries and as the political and cultural center of Mediterranean civilization. Beginning in the 4th century, it became the spiritual center of the Christian world, thus earning the title "Christian Rome". In the 16th century, Pope Sixtus V transformed the city to connect seven major churches and provide roads and directions for pilgrims, establishing a system of radial road axes. The most famous example is People's Plaza, from which three roads radiate, leading to the city's important churches, squares, and ancient Roman landmarks like the Colosseum. At key points along these roads, monumental obelisks and several fountains were erected to offer spatial orientation and visual appeal. These transformations are regarded by European architects as the beginning of Baroque urban planning.

The "Historic Center of Rome" was inscribed on the *World Heritage List* in 1980, encompassing the walled city from the 17th century as well as St. Paul's Cathedral outside the walls. This World Cultural

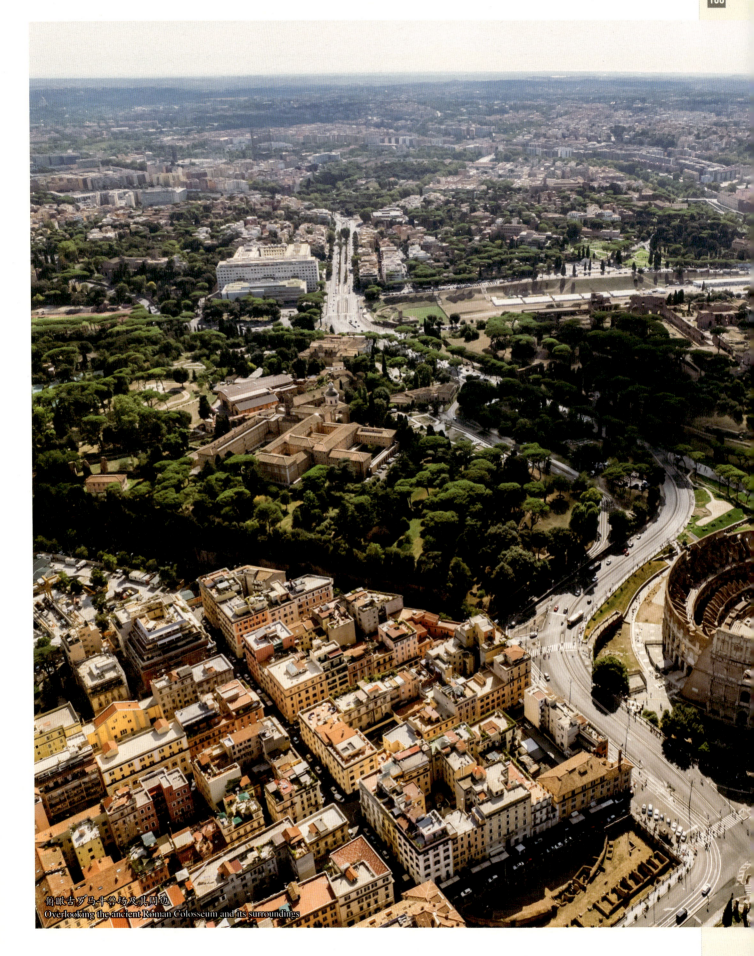

俯瞰古罗马斗兽场及其周边
Overlooking the ancient Roman Colosseum and its surroundings

Heritage site showcases achievements in urban development from the classical period, the Renaissance, and the Baroque period, especially the urban axes established during the Baroque era.

古罗马斗兽场日出
Sunrise at the ancient Roman Colosseum

古罗马广场，古罗马时代的城市中心
The Roman Forum, the urban center of ancient Rome

从圣天使城堡向外眺望圣天使桥
Looking out from Castel Sant'Angelo at the Sant'Angelo Bridge

波波洛广场及三条向外的街道
People's Plaza and the three streets leading outward

中轴视角之意大利罗马
Rome, Italy from the Central Axis Perspective

　　罗马城的改造是宗教意图的展现，在给前来朝拜的信众提供交通便利的同时，也十分重视城市轴线在视觉景观上的表达，塑造出统一的、充满神圣感的城市景观轴线。同时，城市道路呈现多中心的放射形，而轴线是放射形道路网的一部分。这些特点都与北京中轴线那种体现皇权至上和唯一居中主轴的风格大不相同。

　　The transformation of Rome reflects religious intentions, providing convenience for pilgrims while emphasizing the visual expression of urban axes, creating a unified and sacred urban landscape. The city's roads exhibit a multi-centered radial pattern, with the axes forming part of this network. These characteristics contrast sharply with the style of Beijing Central Axis, which embodies supreme imperial authority and a singular central orientation.

纳沃纳广场及广场内的喷泉与方尖碑
Piazza Navona with its fountains and obelisk

圣彼得广场，向东通向圣天使堡
St. Peter's Square, leading east to Castel Sant'Angelo

圣彼得大教堂穹顶
The Dome of St. Peter's Cathedral

西班牙马德里
Madrid, Spain

清晨的马德里丽池公园
Morning at Madrid's Retiro Park

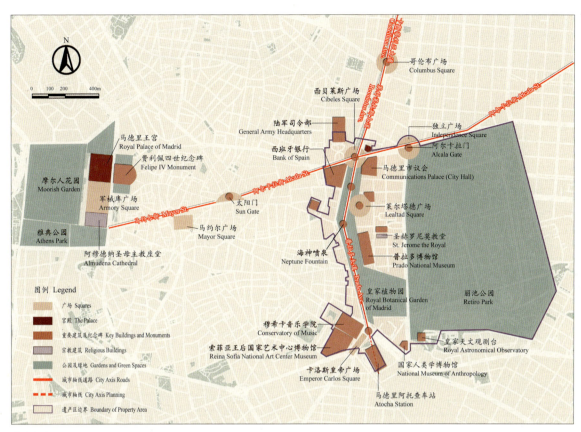

普拉多大道和丽池公园轴线示意图
Diagram of the axis between Prado Avenue and Retiro Park

马德里自 1561 年以来一直是西班牙的首都，而马德里的十字交叉城市轴线的形成也与其首都的历史一样久远。16 世纪末，这里形成了马约尔街、阿尔卡拉街构成的东西向轴线；17 世纪中叶，又形成了卡斯蒂利亚大道、莱科莱托斯大道、普拉多大道构成的南北轴线，两者相交于西贝莱斯广场。

其中普拉多大道原本是一条旧城郊区的休闲步行道，被认为是西班牙林荫大道的原型，园林景观价值卓越。如今，西贝莱斯广场和卡洛斯皇帝广场位于普拉多大道两端，国家政治、经济、科学文化与艺术等公共建筑分列道路两侧，这条南北向的大道既包括了怡人的林荫道，两侧也排列着具有西班牙新古典主义风格和 20 世纪流行的历史主义风格的建筑，自然与人文有机地结合在一起。面积 120 公顷的丽池公园坐落在十字轴线的东南，原本是 17 世纪布恩·丽池宫的遗迹，如今成为 19 世纪至今的欧洲多种园林风格的荟萃之地。

"普拉多大道和丽池公园，艺术与科学的景观之地"于 2021 年以遴选标准(ii)、(iv)、(vi)列入《世界遗产名录》，包含了马德里自文艺复兴和启蒙运动时期发展形成的城市景观。

Madrid has been the capital of Spain since 1561, and the formation of its intersecting city axes is as historical as its status. In the late 16th century, an east-west axis was established by Mayor Street and Alcala Street. By the mid-17th century, the north-south axis emerged, comprising Castellana Avenue, Recoletos Avenue, and Prado Avenue, intersecting at Cibeles Square.

Prado Avenue originally served as a leisure walkway in the outskirts of the old city, considered a prototype of Spain's tree-lined boulevards, with exceptional landscape value. Today, Cibeles Square and Emperor Carlos Square are located at either end of Prado Avenue, flanked by public buildings representing national politics, economy, science, culture, and art. This north-south avenue features pleasant tree-lined paths alongside architecture in the Spanish Neoclassical

独立广场中央的阿尔卡拉门
The Alcala Gate in the center of Independence Square

style and popular 20th century historicism, blending nature and culture seamlessly. Retiro Park, covering 120 hectares, lies to the southeast of the intersecting axes and originated from the 17th century Buen Retiro Palace. It has become a showcase of various European garden styles since the 19th century.

The site "Prado Avenue and Retiro Park: A Landscape of Arts and Sciences" was inscribed as a World Heritage Site in 2021, reflecting Madrid's urban landscape development since the Renaissance and the Enlightenment.

西贝莱斯广场
Cibeles Square

太阳门广场，有10条道路从这里辐射开来
Sun Gate Square, from which ten roads radiate out

阿方索十二世纪念碑，位于丽池公园内
The Alfonso XII Monument, located within Retiro Park

海神喷泉，位于普拉多大道
Neptune Fountain, located on Prado Avenue

丽池公园内的园艺景观
The horticultural landscape within Retiro Park

俄罗斯圣彼得堡
St. Petersburg, Russia

白雪村托中的海军总部大楼金顶
The golden dome of the Admiralty, accentuated by the white snow

圣彼得堡城市轴线示意图
Diagram of the urban axes of St. Petersburg

在夺取波罗的海出海口后，彼得大帝于1703年开始在海边修建一座宏大的城市即圣彼得堡，这座被称为"北方威尼斯"的城市以其无数的河道和400多座桥梁而闻名于世，运河、街道和码头构建了城市交通网络。

18世纪初圣彼得堡的城市轴线开始修建，至1750年左右，形成了以海军总部大楼为顶点，向东、东南和南放射出去的三条街道轴线，这一设计明显借鉴了欧洲巴洛克城市的规划传统。海军总部大楼正立面长107米，正中一座72米的高塔成为视觉焦点；大楼东北侧修建了赫赫有名的冬宫，并由附近若干建筑围成了冬宫广场；大楼西南侧是枢密院大厦，与南侧的圣以撒大教堂合围形成枢密院广场。背依海湾的海军总部大楼与两侧的广场、三条放射道路构成了壮丽的城市空间。

"圣彼得堡历史中心及其相关古迹群"于1990年以遴选标准(i)、(ii)、(iv)、(vi)列入《世界遗产名录》，遗产区范围包括围绕河口形成的历史城区，以及形成于18世纪的城市轴线。

In 1703, after seizing the Baltic Sea outlet, Peter the Great began constructing the grand city of St. Petersburg, known as the "Venice of the North" for its numerous canals and over 400 bridges, which form its urban transport network.

By the early 18th century, the city's axis was established, culminating around 1750 with three radiating streets from the Admiralty. This design clearly drew inspiration from European Baroque urban planning. The Admiralty, with its 107-meter facade and a central tower reaching 72 meters, serves as a visual focal point. Nearby, the famous Winter Palace forms part of the Palace Square, while the Senate Building and St. Isaac's Cathedral create the Senate Square to the southwest. Together, these elements create a magnificent urban space.

The "Historic Center of St. Petersburg and Related Groups of Monuments" was inscribed on the *World Heritage List* in 1990, encompassing the historical district around the river mouth and the 18th century urban axis.

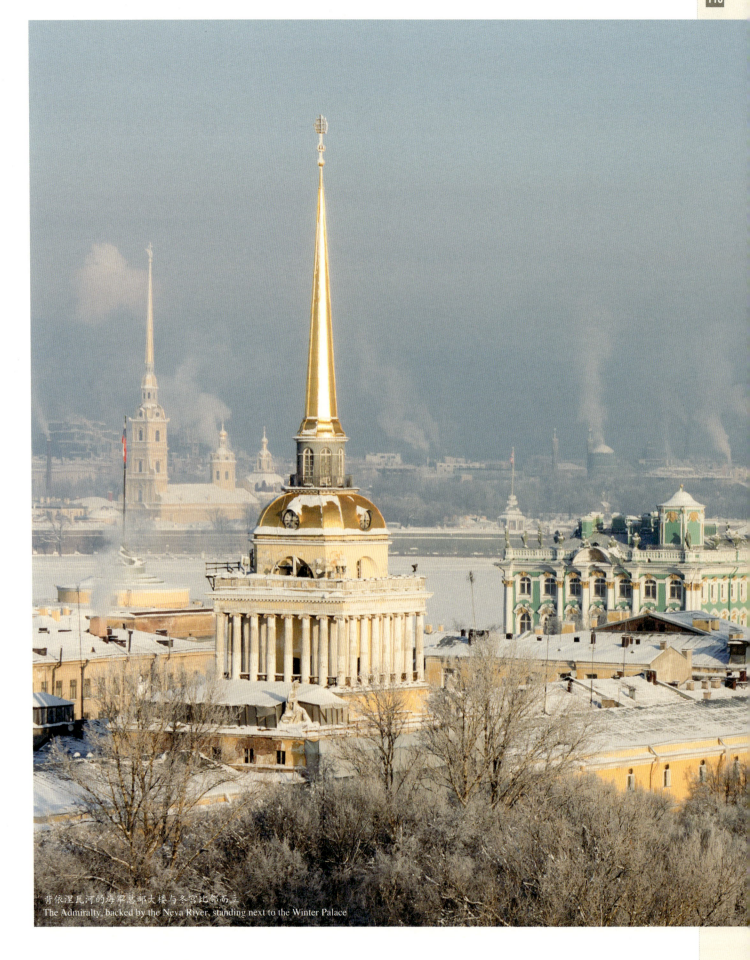

背依涅瓦河的海军总部大楼与冬宫比邻而立
The Admiralty, backed by the Neva River, standing next to the Winter Palace

隔河远望海军总部大楼和圣以撒大教堂
From across the river, the Admiralty and St. Isaac's Cathedral can be seen in the distance

以海军总部大楼为顶点，三条放射形道路通向远方
Three radial roads leading away from the Admiralty as the apex

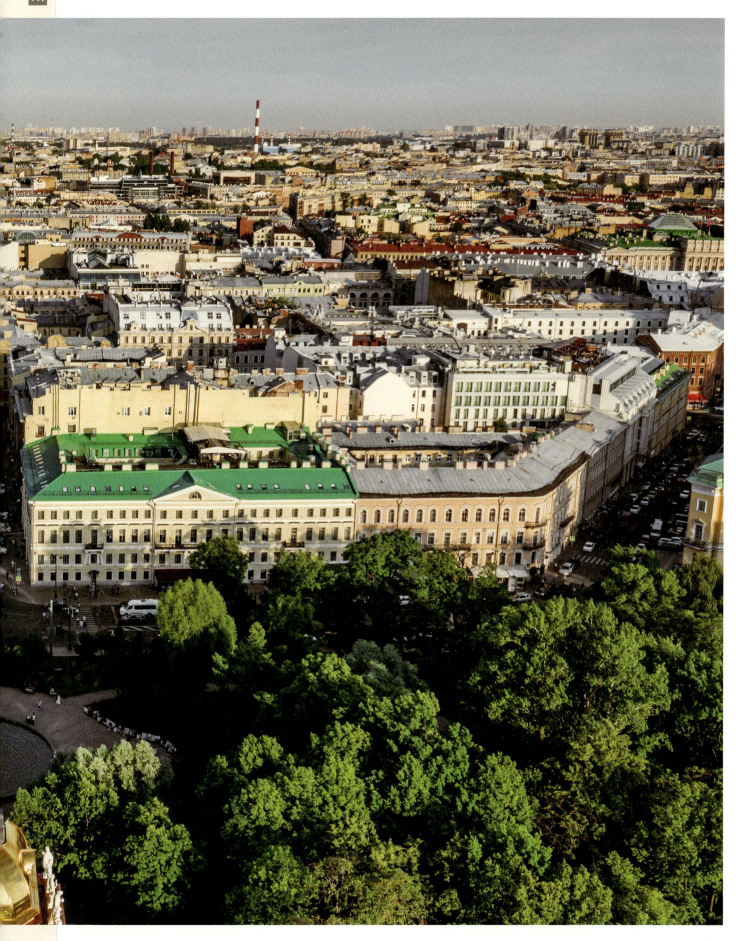

从总参谋部大楼拱门向外看，亚历山大柱高高耸立
Looking out from the arch of the General Staff Building, the Alexander Column rises high

中轴视角之俄罗斯圣彼得堡
St. Petersburg, Russia from the Central Axis Perspective

　　这座都市的营建，代表了古老的俄罗斯向欧洲西部列强学习和开放的心态，所以很多欧洲建筑风格都在这里得到了尝试，使得城市的面貌快速多变。而且，流淌的涅瓦河也给城市的布局带来了空间的可塑性和丰富多样的景观，甚至城市轴线也与河流走向具有关联性。涅瓦河的水上空间与岸上的广场系统衔接在一起，包括轴线在内的街道系统在这样开阔的背景下延展。总之，圣彼得堡的轴线系统与北京中轴线截然不同，二者反映了完全不同的设计理念和文化背景。

冬宫拉斐尔画廊
The Winter Palace's Raphael Gallery

　　This city's construction represents ancient Russia's openness to learning from the Western European powers, leading to a rapid experimentation with various European architectural styles. The flowing Neva River adds spatial flexibility and diverse landscapes to the city's layout, with the city's axis closely related to the river's course. The water space of the Neva connects with the square system on land, allowing the street system, including the axes, to extend within this expansive backdrop. In summary, St. Petersburg's axis system contrasts sharply with Beijing Central Axis. They reflect completely different design concepts and cultural backgrounds.

圣以撒大教堂
St. Isaac's Cathedral

横穿圣彼得堡核心区域的莫伊卡河
Moika River traversing the core area of St. Petersburg

法国斯特拉斯堡

Strasbourg, France

斯特拉斯堡大教堂内部
The Interior of Strasbourg Cathedral

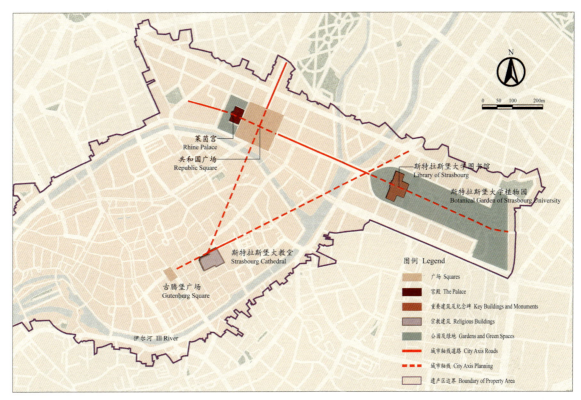

斯特拉斯堡城市轴线示意图
Diagram of the urban axes of Strasbourg

斯特拉斯堡位于法国东端，地处西欧的中心地带，自古以来就是商贸中心。从 19 世纪开始，德国与法国争霸欧陆，位于国境线上的斯特拉斯堡几度易手，这样的历史背景让斯特拉斯堡受到德法两国文化的影响，并体现在城市规划和建筑风格上。

斯特拉斯堡大岛被莱茵河支流伊尔河环绕，它以 15 世纪所建的主教堂为核心，保持了中世纪城镇道路街巷曲径通幽的特点。新城建于 19—20 世纪初，具有十字交叉的轴线，交叉处为共和国广场，是核心行政区所在地，东西向轴线是主轴线，莱茵宫和斯特拉斯堡大学分别位于两端，轴线沿自由大道联系了行政区和大学校园区，两侧配以园林景观；南北向轴线沿和平大道延伸，政府办公大厦分列道路两侧，至高耸的斯特拉斯堡大教堂止。新城的轴线布局借鉴了巴黎的规划，而建筑风格又富有日耳曼特色，体现出法国和德国文化的交融。

斯特拉斯堡大岛于 1988 年列入《世界遗产名录》，遗产区是围绕主教堂的、中世纪发展至今的大岛区域，2017 年遗产地扩展，新增遗产区为 1871—1918 年规划和建造的新城区域。

Strasbourg, located at the east edge of France, sits at the center of Western Europe and has historically been a trade hub. Since the 19th century, as Germany and France vied for the dominance in continental Europe, Strasbourg, positioned on the border, changed hands multiple times. This historical context has led to a blend of German and French cultural influences, evident in its urban planning and architectural styles.

The Grande Île is surrounded by the Ill River, with the 15th century cathedral at its core, preserving the characteristic winding streets of a medieval town. The new city, built from the late 19th to early 20th centuries, features a cross-shaped axis, with Republic Square at the intersection, serving as the core administrative area. The east-west axis is the main thoroughfare, flanked by the Rhine Palace and the Strasbourg University at either end, linking the administrative district to the university

莱茵宫与共和国广场
Rhine Palace and Republic Square

campus along Avenue de la Liberté, enhanced by landscaped gardens. The north-south axis extends along Avenue de la Paix, with government buildings lining on both sides, culminating at the towering Strasbourg Cathedral. The layout of the new city draws inspiration from Parisian planning, while its architectural style exhibits distinct Germanic characteristics, reflecting the fusion of French and German cultures.

In 1988, the Grande Île was designated a UNESCO World Heritage Site, encompassing the medieval area surrounding the cathedral. In 2017, the heritage site was expanded to include the new city area developed from 1871 to 1918.

斯特拉斯堡大教堂
Strasbourg Cathedral

圣皮埃尔·勒·琼天主教堂
Saint-Pierre-le-Jeune Catholic Church

斯特拉斯堡大教堂鸟瞰
Aerial view of Strasbourg Cathedral

莱茵宫与圣皮埃尔·勒·琼天主教堂连线及周边
Connection between Rhine Palace and Saint-Pierre-le-Jeune Catholic Church, along with the surrounding area

126

时代交响

Symphony of the Times

近现代都市轴线

Modern Cities with Prominent
Axial Layouts

樱花掩映中的杰斐逊纪念堂
Thomas Jefferson Memorial amid cherry blossoms

塞内加尔圣路易斯岛
Island of Saint-Louis, Senegal

美国华盛顿特区国会大厦
The Capitol, Washington, D.C., USA

澳大利亚堪培拉国会大厦
Parliament House, Canberra, Australia

阿根廷布宜诺斯艾利斯的马德罗港, 位于东西向轴线的延长线上
Puerto Madero, Buenos Aires, Argentina, located on the extension of the east-west axis

巴西巴西利亚帕拉诺瓦湖和跨湖大桥
Paranoa Lake and the JK Bridge, Brasilia, Brazil

以大航海时代为序幕，16世纪以来，西方殖民主义风潮曾在几百年中席卷各个大洲，许多古代政权纷纷湮灭，代之以西方殖民地。此后，随着殖民地人民独立运动的兴起，世界上涌现出很多新兴国家。在这几百年的风云变幻中，欧洲的城市规划理念被殖民地或新兴独立国家所借鉴和应用，很多国家的都市用较短的时间营建出规划整齐、规模可观的现代城市，轴线运用成为这些城市的普遍特色之一。

无论是在非洲、北美洲、大洋洲、南美洲，都可以找到很多这样的案例。非洲塞内加尔长期作为法国殖民地存在，其圣路易斯岛的城市轴线规划就明显受到了早期法国巴洛克城市规划理念的影响，通过轴线形成有效的城市交通廊道。美国独立后，首都华盛顿的营建同样反映了对当时流行的法国巴洛克城市规划理念的实践。澳大利亚首都堪培拉营建更为晚近，因此城市轴线系统更具现代色彩，以多中心、放射形的城市结构为骨架，以国会、议会大厦、法院等不同功能的政治建筑为轴线节点，象征三权分立等现代主义政治主张，同时以园林植物营造秩序性的景观，实现自然与人文的有机结合。南美洲的阿根廷曾经是西班牙的殖民地，那里的布宜诺斯艾利斯和拉普拉塔的城市轴线受到早期现代主义规划理念的影响，作为行政中心或公共空间景观塑造的基准线；巴西曾经是葡萄牙的殖民地，独立后新建首都巴西利亚时，规划了明显的十字交叉轴线，以轴线控制整个城市的布局，表达建设者对于理想城市生活的追求和理想政治生活的愿景，具有更加典型的现代主义城市规划理念。

由于与东亚相距遥远且文化背景迥异，以上案例中的都市所建立的轴线系统，与北京中轴线在规划理念、规划格局、构成要素、景观形态等方面均差异巨大。近代以来，这些区域经历了殖民主义的冲击与渗透，城市轴线的规划和建设深受欧洲各个时期思潮的影响，并结合当地气候条件和文化特征稍作发挥。即便摆脱宗主国独立之后，这些区域的国家在营建都市时，仍沿袭和借鉴了欧洲的规划理念。相比之下，近代的北京一以贯之地践行着儒家文化背景下的传统哲学理念和美学意念，以都城中轴线表达对礼仪、秩序的追求。因此，从古至今，两者虽然不断演进，但彼此之间的城市面貌和轴线形态与内涵差别很大，各美其美。

With the Age of Exploration as a prologue, Western colonialism swept across continents for centuries starting in the 16th century, leading to the demise of many ancient regimes and their replacement with Western colonies. Subsequently, the rise of independence movements in these colonies gave birth to numerous emerging nations. Throughout this turbulent period, European urban planning concepts were adopted and applied by colonies and newly independent countries, resulting in the rapid construction of well-planned, substantial modern cities, where axial layouts became a common feature.

In Africa, North America, Oceania, and South America, many examples can be found. Senegal, as a French colony for long, showcases the influence of early French Baroque urban planning in the design of the urban axis on the Island of Saint-Louis, which facilitates effective urban transit corridors. After the United States of America gaining independence, Washington, D.C. reflected a similar implementation of popular French Baroque planning concepts in its construction. The capital of Australia, Canberra, was built more recently, featuring a modern axial system characterized by a multi-centered, radiating urban structure. Key political buildings like the Parliament House and courts serve as nodes along these axes, symbolizing the principles of modern democratic governance, while landscaping gardens create an ordered environment that harmonizes nature and culture. Argentina, a former Spanish colony, has Buenos Aires and La Plata with urban axes influenced by early modernist planning, serving as benchmark lines for administrative centers and public spaces. Brazil, once a Portuguese colony, built its new capital, Brasilia, after independence, with a distinct cross-axis layout that governs the city's organization, reflecting the builders' aspirations for an ideal urban and political life, showcasing a more typical modernist urban planning philosophy.

In modern times, these regions have experienced the impacts of colonialism, with urban axis planning heavily influenced by various European ideologies, and adapted to local climatic conditions and cultural characteristics. In contrast, Beijing has consistently adhered to traditional philosophical and aesthetic ideals rooted in Confucian culture, expressing the pursuit of ritual and order through its central axis in the capital.

塞内加尔圣路易斯岛
Island of Saint-Louis, Senegal

圣路易斯岛清真寺尖塔和椰树风情
The minaret of the mosque and the charm of palm trees on the Island of Saint-Louis

圣路易斯岛轴线示意图
Diagram of the axis of the Island of Saint-Louis

　　圣路易斯岛是贴近西非海岸线的一座狭长的岛屿，作为曾经的法国殖民地和法属西非的首都，其城市规划深受法国巴洛克城镇风格的影响。狭长的岛屿分成核心区、北区和南区三部分，政府宫殿和广场组成了核心区。岛上道路形成正交网格状的街道格局，历史建筑主要修建于1720—1820年，为带阳台的法国殖民建筑式样，风格统一。

　　19世纪中期以后，两座大桥将圣路易斯岛与东侧的大陆和西侧的巴巴里长滩联络起来，桥梁与核心区形成东西向的轴线，贯穿了岛上的古典建筑群和广场，坐落在大陆和巴巴里上的艾伯雷韦德广场和德拉肖米埃广场分别位于轴线东西两端。

　　The Island of Saint-Louis is a narrow island located close to the West African coastline. As a former French colony and the capital of French West Africa, its urban planning is heavily influenced by the French Baroque town style. The elongated island is divided into three parts: the core area, the north zone, and the south zone, with the government palace and plaza forming the core area. The roads on the island create an orthogonal grid pattern, and the historic buildings, primarily constructed between 1720 and 1820, are characterized by a unified French colonial architectural style with balconies.

　　Since the mid-19th century, two bridges connected the Island of Saint-Louis to the mainland on the east and to the long beach of Barbarie on the west. These bridges formed an east-west axis with the core area, extending through the classical architectural ensemble and plazas on the island. On the axis, Chaumier Square and Ablayewade Square are located at either end on the mainland and Barbarie, respectively.

圣路易斯岛殖民时期的房屋风格
The colonial-era architectural style of the Island of Saint-Louis

连接圣路易斯岛与非洲大陆的费德赫贝大桥
Faidherbe Bridge connecting the Island of Saint-Louis to the African mainland

中轴视角之塞内加尔圣路易斯岛
The Island of Saint-Louis, Senegal from the Central Axis Perspective

作为一个典型的殖民城市，圣路易斯岛的城市轴线规划来自早期法国奥斯曼巴洛克城市规划理念，轴线形态建设相对简单，以形成有效的城市交通廊道为目标。城市轴线虽然横贯该岛屿，但并未对整个城市形态产生很强的控制力。城市建筑风格以法国殖民风格为主，那些带阳台的房子、如画廊般的房子制造出城市的美感，整个城市趋于风格统一和同质化。所以，这座殖民城市不论是所处的历史时期、规划理念、建筑形态，都具有典型的近代早期欧洲风格，与遥远东方的北京中轴线和北京的文化传统大不一样。

As a typical Western colonial city, the urban axis planning of the Island of Saint-Louis is derived from early French Ottoman Baroque planning concepts. The construction of the axis is relatively simple, aiming to create effective urban transit corridors. Although the urban axis traverses the island, it does not exert strong control over the overall city form. The architectural style is predominantly French colonial, with balconied houses and gallery-like buildings contributing to the aesthetic appeal of the city, resulting in a unified and homogeneous style. Thus, this colonial city reflects the early modern European style in its historical context, planning principles, and architectural forms, which starkly contrasts with Beijing Central Axis and the cultural traditions of Beijing.

美国华盛顿

Washington, D.C., USA

从国会山俯瞰东西向轴线绿地和华盛顿纪念碑

The Aerial View of the east-west axis of green space and the Washington Monument from Capitol Hill

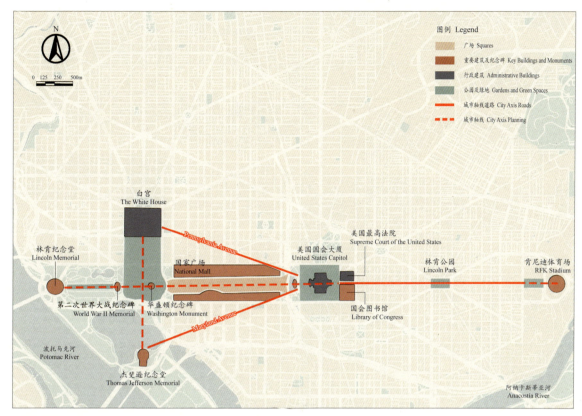

华盛顿城市轴线示意图
Diagram of the urban axes of Washington, D.C.

美国独立后，南方州与北方州就国家的首都位置发生争执，最终双方妥协，在当时南方州靠近北方的位置建设一座新的首都，这就是华盛顿。

1791 年，在法国建筑师的规划下，城市选址在波托马克河和阿纳卡斯蒂亚河交汇处，以区域内高出河面 30 米的詹肯山为国会大厦的位置，以此为基点确定了东西向的城市主轴线。从国会大厦出发，轴线向西通达波托马克河边，向东遥指阿纳卡斯蒂亚河。第二年奠基的白宫及其附属绿地从南北向贯穿东西主轴线，形成了南北向的轴线。

进入 20 世纪，美国政府又对华盛顿轴线区域进行扩展，东西轴线向东延伸，经过林肯公园到达肯尼迪体育场，终于抵达阿纳卡斯蒂亚河畔。而白宫轴线向南区域改造为一池碧水，与波托马克河连通起来。

最终，华盛顿的城市轴线形成了典型的十字结构。轴线规模宏大，东西轴线总长度达 7 千米，南北长度为 1.7 千米，以东西轴线为主轴，以南北轴线为副轴，分别象征着立法与行政的独立以及三权分立的政治体制；城市轴线因中央绿化带而凸显其结构，轴线的视觉节点上放置了华盛顿纪念碑等建筑，轴线两侧分布着大量公共建筑，如美术馆、博物馆、纪念堂等。

华盛顿的营建，反映了法国巴洛克式轴线规划对北美的影响，并展现出 20 世纪以来城市美化运动的风格。

Since the United States gained independence, the southern and northern states have disagreed over the site of the nation's capital. Ultimately, a compromise was reached to build a new capital on the site within the southern states but near the north, which was known as Washington, D.C.

国会大厦和两边博物馆林立的东西向绿地轴线
The Capitol Building and the east-west green axis lined with museums

In 1791, under the planning of a French architect, the city was situated at the confluence of the Potomac and Anacostia Rivers. Jenkins Hill, 30 meters higher than the river level, was chosen as the location for the Capitol Building, serving as the starting point for the city's east-west main axis. This axis extends westward to the Potomac River and eastward toward the Anacostia River. The White House, with its surrounding green spaces, was established the following year and cuts across the east-west axis, forming a north-south axis.

In the 20th century, the U.S. government expanded the axis region. The east-west axis extended eastward, passing through Lincoln Park, reaching RFK Stadium and ultimately arriving at the Anacostia River. The south of the White House was transformed into a pool, connecting to the Potomac River.

Eventually, Washington's urban axis formed a typical cross structure. The scale is grand, with the east-west axis extending 7 kilometers, and the north-south axis reaching 1.7 kilometers. The main axis is the east-west line, while the north-south axis serves as a secondary line, symbolizing the independence of the legislative and administrative branches and the separation of powers. The structure of the axis is highlighted by a central greenbelt, with visual nodes such as the Washington Monument placed along the axis. Numerous public buildings, including art galleries, museums, and memorials, are located along both sides of the axis.

The construction of Washington reflects the influence of French Baroque axial planning on North America and showcases the style of the City Beautiful movement that emerged in the 20th century.

国会大厦背侧
The rear view of the Capitol Building

最高法院大楼
Supreme Court Building

国会图书馆
Library of Congress

林肯纪念堂，位于东西向轴线西端
Lincoln Memorial, located at the west end of the east-west axis

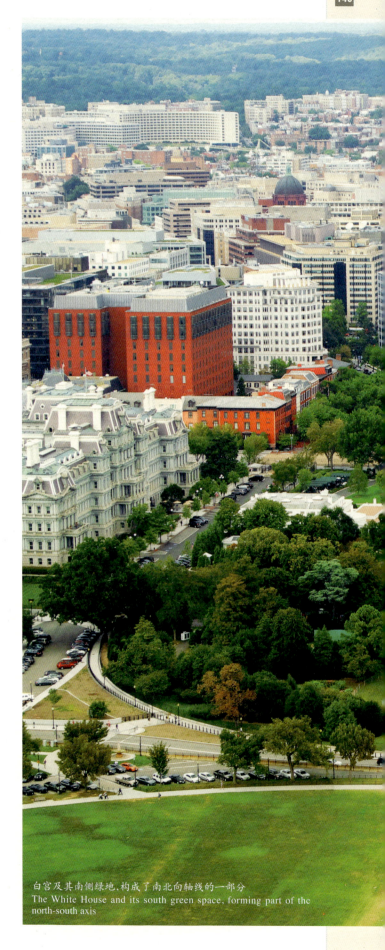

白宫及其南侧绿地，构成了南北向轴线的一部分
The White House and its south green space, forming part of the north-south axis

澳大利亚堪培拉

Canberra, Australia

与格里芬湖比邻的澳大利亚国家图书馆
The National Library of Australia, adjacent to Lake Griffin

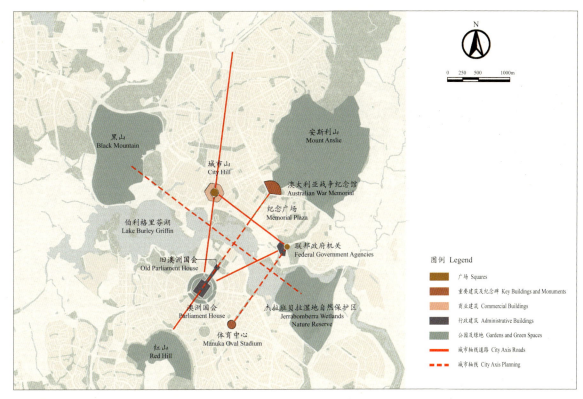

堪培拉城市轴线示意图
Diagram of the urban axes of Canberra

　　堪培拉的兴建与华盛顿确定城址的剧情如出一辙，20 世纪初，澳大利亚联邦政府在选择悉尼还是墨尔本为首都的问题上举棋不定，最终折中选择在两城之间另建新城。

　　堪培拉的规划深受当时田园城市思想的影响，它以国会大厦为城市核心，大致向北、东、南三个方向延伸出数条放射形道路，并围绕中心形成同心圆形的城市布局。政府部门和文化机构主要分布在中心向北和向东北两条轴线之间的扇形区域，其他方向则分布着商业区和生活区等。从国会大厦出发的近北方向的轴线跨河与北岸的城市山相联系，而城市山则是另一条放射形道路与同心圆布局的中心，于是两个中心的连线就成为堪培拉的主轴线。

　　主城区两侧的山峦和大体上东西展布的湖泊、河流与主轴线错落有致地排布在一起，形成一幅都市与自然融合的美景。

The construction of Canberra mirrors the narrative of determining the capital site in Washington, D.C. In the early 20th century, the Australian federal government was indecisive about whether to choose Sydney or Melbourne as the capital, ultimately opting to build a new city between the two.

Canberra's planning was heavily influenced by the concept of the garden city, with the Parliament House as its core. Several radial roads extend roughly north, east, and south from this center, creating a concentric circular layout. Government departments and cultural institutions are mainly distributed within a fan-shaped area between the north and northeast axes, while other directions contain commercial and residential zones.

The axis extending north from the Parliament House crosses the river to connect with the City Hill, which serves as another focal point in the radial layout. Thus, the line connecting these two centers forms Canberra's main axis. The surrounding mountains and the lakes and rivers that generally spread east-west harmoniously merged with the main axis, creating a picturesque blend of urban and natural landscapes.

澳大利亚战争纪念馆，位于澳大利亚国会东北方向，隔格里芬湖相望
Australian War Memorial, located northeast of the Australian Parliament and facing it across Lake Griffin

城市山，国会之外的另一个放射形中心
City Hill, another radial center outside of the Parliament

前方建筑为旧国会大厦，后方建筑为国会大厦，是堪培拉的城市核心
In the foreground is the Old Parliament House, while the building behind it is the Parliament House, forming the urban core of Canberra

从安斯利山高点向西南眺望国会大厦
Overlooking southwest the Parliament House from the summit of Mount Anslie

从国会大厦向东北方向延伸的国王大道,远方是澳大利亚-美利坚纪念碑
Kings Avenue extending northeast from the Parliament House, with the Australian-American Memorial in the distance

中心圆明显的堪培拉城市核心区
The central circle distinctly marking Canberra's urban core area

阿根廷布宜诺斯艾利斯和拉普拉塔

Buenos Aires and La Plata, Argentina

布宜诺斯艾利斯圣马丁广场上的纪念钟楼，位于南北向轴线的北方
The memorial clock tower at Plaza San Martín in Buenos Aires, located at the northtrend of the north-south axis

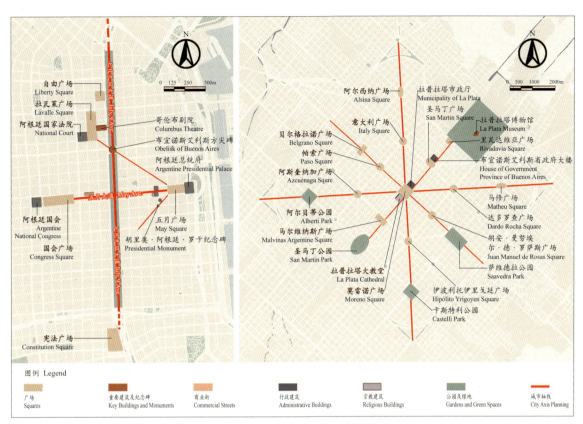

布宜诺斯艾利斯城市轴线示意图
Diagram of the urban axes of Buenos Aires

拉普拉塔城市轴线示意图
Diagram of the urban axes of La Plata

布宜诺斯艾利斯和拉普拉塔两座城市都位于阿根廷拉普拉塔河畔的潘帕斯平原上,前者设计了十字交叉的道路网络和轴线系统,后者规划了对角线道路与规整的道路网络交错,别具一格。

布宜诺斯艾利斯作为阿根廷首都,设计有一条东西方向的五月大道,东端为总统府,面对西面的五月广场,广场西端分别向西北和东南方向延伸出两条大道,广场周边和放射形大道两侧分布着重要的行政机关以及纪念碑等建筑。与五月大道相交的南北方向轴线名为七月九日大道,是以法国香榭丽舍大街为原型的一条林荫大道,沿途设有多处广场、纪念性建筑、剧院等公共建筑。

拉普拉塔城市规整,如同旋转45°的正方形,城市路网由多个6×6方形网格组成,另有正南北和正东西两条对角线道路与之交叉,从城市中心的广场出发,向四面八方有8条放射形道路,城市结构具有较强的向心性。

两座城市建筑风格多样,展现出欧洲文化和南美地方风格结合所形成的折中主义文化。

Buenos Aires and La Plata are both located on the Pampas plain along the Río de la Plata in Argentina. Buenos Aires features a cross-shaped road network and axis system, while La Plata incorporates diagonal roads interwoven with a structured road network, giving it a unique character.

As the capital of Argentina, Buenos Aires is designed around May Avenue, which runs east-west. At its east end is the presidential palace facing May Square, from which two avenues extend northwest and southeast. Important administrative buildings and monuments are distributed around the square and along the radial avenues. The north-south axis intersecting May Avenue is July 9 Avenue, a tree-lined boulevard modeled after Paris's Champs-Élysées, featuring numerous squares, monumental buildings, theaters, and other public buildings.

布宜诺斯艾利斯五月广场
May Square, Buenos Aires

La Plata is laid out in a neat square rotated 45 degrees, with a grid of 6×6 squares and additional north-south and east-west diagonal roads. From the central square, eight radial roads extend outward, creating a strong centripetal structure.

Both cities exhibit diverse architectural styles, showcasing a blend of European culture and local South American influences, resulting in an eclectic cultural identity.

阿根廷总统府，位于五月广场东侧
Argentine Presidential Palace, located on the east side of May Square

阿根廷国会广场及国会大楼
Congress Square and the Congress Building, Argentina

布宜诺斯艾利斯国会广场
Congress Square, Buenos Aires

拉普拉塔主教教堂，位于城市中心莫雷诺广场西南侧
La Plata Cathedral, located southwest of Moreno Square in the city center

拉普拉塔博物馆，位于菱形城市中心区的东北边侧
La Plata Museum, situated on the northeast side of the diamond-shaped city center

巴西巴西利亚

Brasília, Brazil

巴西利亚东西方向的纪念轴大街
Monumental Axis in Brasília, running east-west

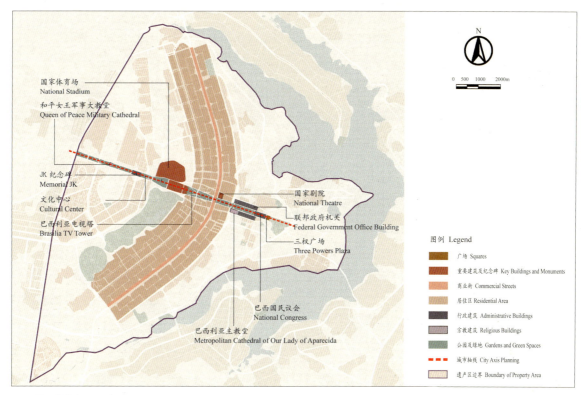

国家体育场
National Stadium

和平女王军事大教堂
Queen of Peace Military Cathedral

JK 纪念碑
Memorial JK

文化中心
Cultural Center

巴西利亚电视塔
Brasília TV Tower

国家剧院
National Theatre

联邦政府机关
Federal Government Office Building

三权广场
Three Powers Plaza

巴西国民议会
National Congress

巴西利亚主教堂
Metropolitan Cathedral of Our Lady of Aparecida

图例 Legend

广场 Squares

重要建筑及纪念碑 Key Buildings and Monuments

商业街 Commercial Streets

居住区 Residential Area

行政建筑 Administrative Buildings

宗教建筑 Religious Buildings

公园及绿地 Gardens and Green Spaces

城市轴线 City Axis Planning

遗产区边界 Boundary of Property Area

巴西利亚城市轴线示意图
Diagram of the urban axes of Brasilia

　　1956 年巴西决定在内陆兴建一座新的首都——巴西利亚，四年后一座现代主义新城拔地而起。人们形容巴西利亚如同张开双翼的飞鸟或飞机，它有一条近东西向的轴线和一条南北向轴线交叉于城市中，构成十字形的城市结构，南北向轴线因背靠帕拉诺瓦湖的自然形态而呈弧形，好似飞翼。

　　东西向轴线长约 6 千米，自西向东沿一条 200 米宽的林荫道和广场延展，依次分布着天主教堂、布里蒂宫、国家体育场、文化中心、电视塔、国家剧院、巴西利亚主教堂、联邦政府机关大楼、巴西国民议会及三权广场。此轴线西端为交通枢纽站，东端到三权广场止，议会大厦、最高法院和普拉纳托宫（总统办公地）占据三角形广场的三个顶点，象征三权分立的理念。南北向轴线两侧张开各约 5 千米，以一条宽阔的快速道路为主干，道路两旁则规划了方形居住单元组成的连续居住街区。

　　巴西利亚于 1987 年以遴选标准(i)、(iv)列入《世界遗产名录》，遗产区范围包含了十字交叉的城市轴线。这座城市鲜明地反映了 20 世纪现代功能城市设计和现代主义城市规划理念，建筑也颇具现代风格，堪称世界城市规划史上的里程碑。

In 1956, Brazil decided to build a new capital in the interior of the country, Brasilia. Four years later, a modernist city rose from the ground. Brasilia is often described as a bird or airplane with outstretched wings, featuring an east-west axis and a north-south axis that intersect within the city, forming a cross-shaped urban structure. The north-south axis follows a curved line, shaped by the natural contours of Paranoa Lake, resembling wings.

The east-west axis is approximately 6 kilometers long, extending from west to east along a 200-meter-wide tree-lined avenue and plaza. Along this axis, the Cathedral, Buriti Palace, National Stadium, Cultural Center, TV Tower, National Theater, Metropolitan Cathedral of Our Lady of Aparecida, Federal Government Office Buildings, National Congress, and Three Powers Plaza are distributed. The west end serves as a transportation hub, while the east end leads

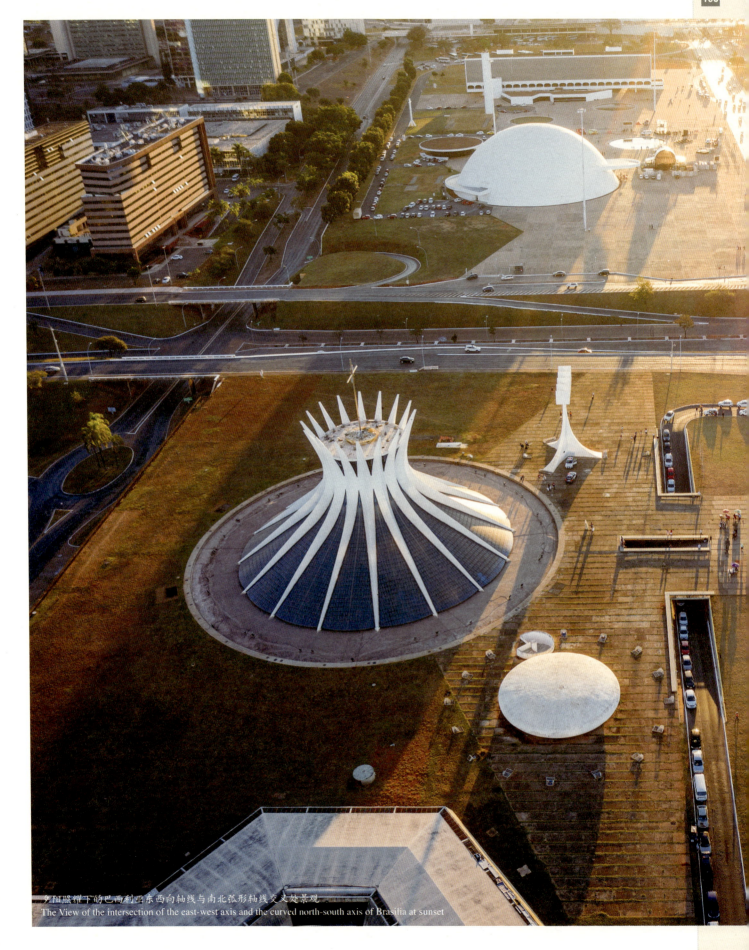

夕阳照耀下的巴西利亚东西向轴线与南北弧形轴线交叉处景观
The View of the intersection of the east-west axis and the curved north-south axis of Brasilia at sunset

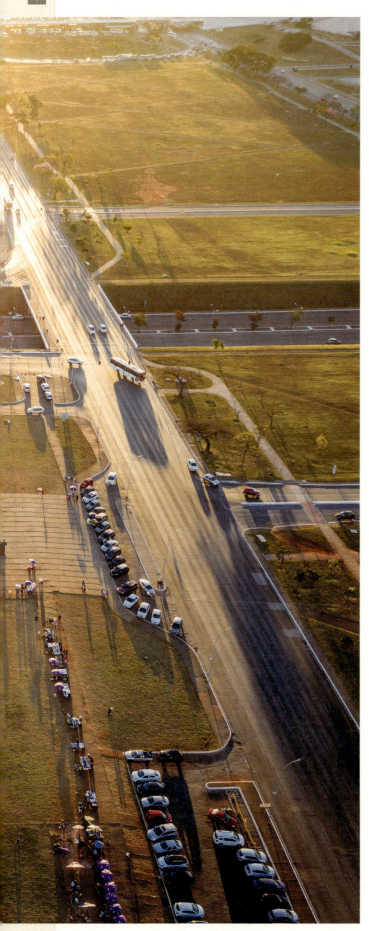

to the Three Powers Plaza, where the Congress, Supreme Court, and Palacio do Planalto (the presidential office) occupy the apex of the triangular plaza, symbolizing the separation of powers. On either side of the north-south axis, approximately 5 kilometers are developed with a broad expressway as the main thoroughfare, flanked by continuous residential blocks composed of square housing units.

In 1987, Brasilia was inscribed as a World Heritage Site, encompassing the cross-shaped urban axes. The city distinctly reflects 20th century modern functional urban design and modernist planning concepts, with architecture that embodies modern style, marking it as a milestone in the history of urban planning worldwide.

巴西利亚主教教堂
Metropolian Cathedral of Our Lady of Aparecida

巴西国民议会大楼，位于东西向轴线的东部
National Congress of Brazil, located on the east side of the east-west axis

巴西最高法院大楼，位于东西向轴线的东端
Supreme Court of Brazil, situated at the east end of the east-west axis

伊塔马拉第宫，位于三权广场附近
Itamaraty Palace, near the Three Powers Plaza

祖国与自由万神殿，位于三权广场
Pantheon of the Fatherland and Freedom, located at the Three Powers
Plaza

中轴视角之巴西巴西利亚
Brasilia, Brazil from the Central Axis Perspective

　　与当今世界其他都市相比，巴西利亚是一座非常晚近的城市，因此从营建理念上更具现代主义色彩，并且与当地的热带环境进行了有机结合。城市的规划者希望一切设计元素都能够与城市的整体设计相吻合。同时，这座都市也体现了现代民主制度的一些追求，比如三权分立；城市轴线和建筑的设计中融入了大量几何结构的设计。这样一座现代新型都市与古朴典雅的北京中轴线和古都北京有着巨大的反差。

Compared to other cities around the world, Brasilia is a relatively recent city, characterized by modernist principles that harmonize with the local tropical environment. The city planners aimed for all design elements to align with the overall urban design. Additionally, Brasilia reflects aspirations of modern democratic ideals, such as the separation of powers, with a significant incorporation of geometric structures in its axis and architectural designs. This modern urban landscape contrasts sharply with the traditional elegance of Beijing Central Axis and its ancient cultural heritage.

巴西利亚近南北方向的轴线大街
The major avenue running close to the north-south axis in Brasilia

后记
Afterword

"轴线（Axis）可能是人类最早的现象，这是人类一切行为的方式。刚刚会走的孩子也倾向于按轴线走……轴线是建筑中的秩序维持者……轴线是一条引导目标的线……"

在这段话中，现代主义建筑大师勒·柯布西耶从孩子的视角来理解轴线。也许我们可以就此推论，认知和运用轴线可能是人的一种本性。如果这个想法是合理的，那么古往今来世界各地大量都市都或多或少采用轴线规划全局或局部，也就容易解释了，因为"人同此心，心同此理"。

然而，人类的创造力却又是飘逸无涯的，同样的轴线，在古人与今人、西方人与东方人的不同群体手中，会展现不一样的外观与内涵。于是乎，古与今、西与东的都市轴线就有了差异，有了文明互鉴的基础。

中国古代都城中轴线规划格局来源于古老的"以中为尊"的文化思想，又为礼制所催生的《考工记》所描绘的理想都城范式所引领，从远古一路走来，历经邺北城、洛阳城、东京城、上都……不断扬弃、不断精进，最终到北京中轴线，它作为传统都市轴线系统的集大成者，以中轴线系统串联起皇家宫苑、礼仪祭祀、国家庆典等重要空间，以及城门、钟鼓楼、商业街市等城市管理与服务设施，以恢宏的规模、层进的序列、严谨的景观，彰显传统都城的独特美学，将理想中的都市范式在人间实现。

北京中轴线的规划理念甚至在这条轴线诞生之前，就已经开始影响周边的东亚、东南亚地区，给一衣带水的邻邦都城建设提供借鉴，显示出北京中轴线超越国界的影响力。

近现代以来，在文艺复兴、科学革命、工业革命的助力下，欧洲建筑理念更新迭代，在全球掀起浪潮，从欧洲的巴黎、罗马到圣彼得堡、斯特拉斯堡，从北美洲的华盛顿到南美洲的布宜诺斯艾利斯，从大洋洲的堪培拉到非洲的圣路易斯岛，结构清晰的轴线系统几乎成为近现代新建都市里的标准配置，那刻在人类孩童体内的"轴线基因"已然觉醒并爆发。

在世界数百年的西方建筑浪潮中，很多古城消失或被改造，东方的北京中轴线幸免于难，成为东亚地区现存规模最大且保存最为完好的传统都城中轴线实例，其所承载的古代哲学、古代美学、礼仪秩序在现代化和全球化的时代，更显独特性与唯一性。

2024年7月27日，在印度新德里举行的联合国教科文组织第46届世界遗产大会上，"北京中轴线——中国理想都城秩序的杰作"被正式列入《世界遗产名录》，成为中国第59项世界遗产。

北京中轴线也在与时俱进，融合古老与现代，传统与时尚，它将与世界上的其他都市一道，各美其美，美美与共，共同向着更为美好的未来演进。

"Axis may be one of the earliest phenomena of humanity, a way of all human actions. Even toddlers tend to walk along an axis... An axis is the maintainer of order in architecture... An axis is a line that guides towards a goal."

In this passage, modernist architect Le Corbusier interprets the axis from a child's perspective. We might speculate that the cognition and application of the axis could be a part of human nature. If this idea holds, then the widespread adoption of axial planning in cities worldwide throughout history is understandable, as "people share the same heart, and the heart understands the same principles".

However, human creativity is boundless. The same axis will exhibit different appearances and meanings in the hands of different cultures, whether ancient or modern, East or West. Thus, the differences in urban axes between ancient and modern, East and West, lay the groundwork for mutual learning among civilizations.

The axial planning of ancient Chinese capitals stems from the age-old cultural concept of "centrality", guided by the ideal city model depicted in *Kaogongji*, shaped by ritual systems. From ancient times through cities like Yebei, Luoyang, Dongjing, and Xanadu, it has continuously evolved, culminating in Beijing Central Axis. This axis system connects important spaces such as imperial palaces, ritual sites, and national celebrations, alongside city gates, bell and drum towers, and commercial streets, showcasing a unique aesthetic of traditional capitals.

The planning principles of Beijing Central Axis began influencing neighboring East Asian and Southeast Asian cities even before Beijing Central Axis was built, with its conception demonstrating its transcendent impact beyond borders.

Since the Renaissance, the scientific revolution, and the industrial revolution, European architectural concepts have evolved, sparking a global wave. From Paris, Rome, and St. Petersburg to Washington, D.C. and Buenos Aires, and from Canberra to Saint-Louis in Africa, structured axial systems have become standard in newly built modern cities, awakening the "axis gene" inherent in humanity.

Amidst centuries of Western architectural trends, many ancient cities have disappeared or transformed. In contrast, Beijing Central Axis in the East was fortunate to survive, remaining the largest and best-preserved traditional capital axis in East Asia, carrying ancient philosophy, aesthetics, and ritual order that highlight its uniqueness in the modern global era.

On July 27, 2024, at the 46th session of the World Heritage Committee in New Delhi, "Beijing Central Axis: A Building Ensemble Exhibiting the Ideal Order of the Chinese Capital" was officially inscribed on the *World Heritage List*, which is the 59th World Heritage in China.

Beijing Central Axis continues to evolve, integrating the ancient with the modern and the traditional with the contemporary, joining other cities worldwide in a shared pursuit of a better future.

图书在版编目 (CIP) 数据

各美其美：北京中轴线与世界都市轴线对比：汉、
英 / 吕舟主编 . -- 北京：北京出版社，2025. 6.
ISBN 978-7-200-19400-5

Ⅰ. TU984-49

中国国家版本馆 CIP 数据核字第 2025YZ4768 号

项目策划：刘 扬   责任编辑：高 琪
特约编辑：王宇彤   责任营销：王绍君
装帧设计：王 勇   责任印制：燕雨萌

各美其美

北京中轴线与世界都市轴线对比（汉、英）

GE MEI QI MEI

吕舟 主编

*

北 京 出 版 集 团
北 京 出 版 社   出版

（北京北三环中路 6 号）

邮政编码：100120

网址：www . bph . com . cn

北京伦洋图书出版有限公司发行

河北鑫玉鸿程印刷有限公司印刷

*

889 毫米 ×1194 毫米   16 开本   11.5 印张   235 千字

2025 年 6 月第 1 版   2025 年 6 月第 1 次印刷

ISBN 978-7-200-19400-5

定价：998.00 元

如有印装质量问题，由本社负责调换

质量监督电话：010-58572393